ENCYCLOPAEDIA OF PLANT ECOLOGY

ANMOL PUBLICATIONS PVT. LTD.
H.O.: 4374/4B, Ansari Road, Darya Ganj,
New Delhi-110 002 (India)
Ph.: 23278000, 23261597

B.O.: No. 1015, Ist Main Road, BSK IIIrd Stage
IIIrd Phase, IIIrd Block
Bangalore - 560 085 (India)
Visit us at: www.anmolpublications.com

Encyclopaedia of Plant Ecology

First Published, 2007

ISBN 81-261-3080-6 (Set)

PRINTED IN INDIA

Printed at Mehra Offset Press, Delhi.

Encyclopaedia of
Plant Ecology

Volume 2

A.Z. Bukhari

ANMOL PUBLICATIONS PVT. LTD.
NEW DELHI - 110 002 (INDIA)

Contents

Preface *vii*

1. **Structure of Climate** **1**
Potential of Redox • Assessment of Gravimeter • Titrations Based on Acid • Atmospheric Carcanogens • Elements in Atmosphere • Making of Atmosphere • Factors of Pollution

2. **Plant Protection** **45**
Importance of Plant Ailments • Classification of Plant Ailments • Different Diseases of Plant

3. **Role of Insects** **77**
The Background • Supportive Plants • The Awareness • Practical Aspects • Pest Management • Branches of Zoology • Branches of Entomology • Entomology in Application • Insects and their Structure • Hands at Work • Kinds of Dusters • Sprayers and Kinds

4. **Resources of Earth** **111**
Preserving Soil • Preserving Water • Preserving Energy • Saving Earth • Biosphere • Role of Hydrosphere • Role of Lithosphere • Role of Atmosphere • Complex Cycles

5. **Basic Resources** 139
The Theory • The Classification • The Threat • The Harvesting • Energy in Power • Different Sources • Wealth under Earth

6. **Energy Sources** 157
Different Energy Sources • Greatest Source • Atomic Sides

7. **Nature at Work** 173
Theory of Vitamins • Milch Nature: Breeds • Dual Target of Breeds • Drought Nature of Breeds • New Breeds: Emergence • Use of Domestic Buffalo • Murrah Group's Breed • Breed of Gujarat Group • Buffalo Milk • Natural Resources • Various Means: Forest • Benefits of Jungles • Means of Water • Means of Minerals • Green Resources • Areas Covered • The Framework • The Part • The Flora • The Preservation • Greenery Secured • Trees under Safety • Big Initiative

8. **Human Factors** 231
Growth of Population • Features of Population • Concern of Human Rights • Rehabilitation Programme • Change of Climate • Depletion of Global Warming

9. **Social Factors** 275
Indian Scenario • Saving of Water • Conservation of Rainwater • Environmental Hazards

Bibliography 299

Index 305

Preface

Ecology is the study of relations between plants and living creatures and the surrounding environment. This relation is vital in itself and indicates how significant and necessary plants and animals are for each other. Biologically, each fulfills another's demand for life. This is a multi-faceted issue.

Animals' respiration is based on two types of gases: Oxygen and Carbon dioxide. When we breath in, we take Oxygen inside, released by plants. On the contrary, when we breathe out, we release Carbon dioxide, which is absorbed by plants. In other words, no oxygen, no animal life; no carbon dioxide, no plants.

Plants also serve as an important source of energy for animals in the form of food or fuel. Similarly, animal's excreta is used in fusing new vitality into soil to help seeds grow into plants and trees. Thus, both constitute a never-ending cycle. Besides, there are various other elements, which play important roles in creating a friendly environment for animals, as well as plants.

In fact, it is a vast knowledge which is wellnigh impossible to discuss here. A detailed and research-based study of the subject has been attempted in the form of this multi-volume work, namely, *Encyclopaedia of Plant Ecology*. Researchers students, scholars, academics and the general readers would expectedly find it a useful tool of information in their possession.

We await and admire comments and remarks of the intelligent readers.

— Editor

1

Structure of Climate

Information about the prevailing climate is very important in deciding the ecosystem of a particular area. With the exception of those in the deep ocean, ecosystems are all at least partly shaped by climate.

Climate: Prevailing weather conditions at a location, e.g., temperature, humidity, wind speed and direction, cloud cover and rainfall.

Climate is the outcome of a number of factors:

1. Variations in the amount of incoming solar radiation.
2. The earth's daily rotation and its path around the sun.
3. Distribution of continents and oceans.
4. Elevation of land masses.

These factors interact to produce prevailing winds, ocean currents, and seasonal patterns of temperature and rainfall.

Global air currents influence patterns of temperature and rainfall by the following mechanism.

1. Sun's rays are more concentrated on the equator-as air near the equator heats up, it picks up moisture and begins to rise.

2. As the moist air rises, it cools and releases much of its moisture as rain-high rainfall in the tropics.
3. This drier air moves north and south away from the equator-begins to descend at about 30 degrees north and south latitude and is heated adiabatically as it descends by increasing atmospheric pressure-these descending masses of hot, dry air produce deserts at these latitudes.
4. Some of the air from these descending air masses is diverted back towards the equator, while part continues to move away from equator.
5. Air masses pick up moisture and are heated at 60 degrees latitude-rise and release moisture as rain, e.g., moist climate of Pacific northwest.

Some of this air moves south and some moves north, where it cools and descends on the polar regions-this is dry, cold air and polar regions are actually deserts. The movement of these air masses away from the equator are also influenced by the earth's rotation and the formation of high and low pressure centers over continents and oceans-these factors interact to produce the directions of regional prevailing winds.

Prevailing winds, the earth's rotation and variations in ocean water temperature produce ocean currents which also have distinct regional patterns.

The topography of continents (e.g., mountains, valleys, etc.) influence regional climatic patterns-a rain shadow effect.

Various types of ecosystems found in the world can also be classified. The most basic grouping is based on whether or not the ecosystem is on land (terrestrial) or in the water (aquatic).

Terrestrial ecosystems–regional ecosystem types having similar communities are sometimes called biomes. Biomes arise through a process called succession, i.e. changes in community composition through time.

Primary succession-complete community development from a bare rock surface-involves formation of soil from rock and may take thousands of years to complete-e.g., lava flow to tropical rainforest, a bare area generated after landslide.

Secondary succession-In it existing plant community is removed by a disturbance (e.g., fire)-doesn't include soil formation-time scale is 10-100 years.

In succession, the first community is called the pioneer community and it generally consists of lichens, mosses, weeds, etc. which are able to colonise disturbed areas-gradually each plant community modifies the environment, making it unsuitable for itself, but suitable for another community. In this process gradually each community is replaced by another-process culminates when they reach the climax community which is the final stage in succession.

Climax Communities: persistent, stable collections of species that typify particular regions of the world-climax communities are called biomes by ecologists.

Characteristics of biomes are determined by regional climate and by the types of soils available for plants to grow on.

Examples of some of the major biomes on earth are:

1. Tundra-cold temperatures for much of year-short-growing season (60 days)-may be alpine or arctic-low productivity because of short-growing system-decomposition slow because of cold temperatures-plant litter accumulates and soils have high organic content-low plant species diversity-only grasses, dwarf shrubs grow.
2. Forests-tremendous above-ground biomass-about 40 times more' as productive than tundra-competition for light important to plants-only 1% of sunlight hitting canopy reaches forest floor—forests can be subdivided into:

A. Coniferous (evergreen) forests-cool, wet climates-pines, firs, spruces are common-found in northern latitudes.

B. Temperate deciduous forest-cold regions-e.g., beech, maple, oak, hickory-deciduous trees lose and regrow leaves yearly.

C. Tropical rainforest-most productive of all ecosystems-moderate temperatures and high rainfall (e.g., > 200 cm. per year)-decomposition rates are high-low organic matter in soil-soils have low fertility because nutrients are tied up in tremendous biomass of forest-highest plant and animals species diversity of any biome-e.g., a single acre may contain over 400 different species of trees-rainforests are being destroyed and an estimated 1 species/hour is going extinct due to the anthropogenic disturbances of rainforest.

3. Grasslands-10"-30" rainfall/year-not enough rainfall to support trees-droughts are common-soils very fertile and high in organic matter-grasses are dominant plants and animal community dominated by grazers-prairie of North America-once supported 65 million bison and probably an equal number of pronghorn antelope.

4. Savannas—essentially grassland with scattered trees and shrubs-Africa, Australia and South America-animals consist of grazers (e.g., zebras, rhinos, antelopes) and browsers (elephants, giraffes)-trees deeply rooted and act as nutrient pumps-leaves fall off trees and decompose to make nutrients available to grasses-trees are being used for fuel by humans which is drastically decreasing soil productivity.

5. Shrublands-dominated by shrubs 15'-20' tall-example is the chaparral of southern California-found around Mediterranean and west coast of South America-

common feature is prolonged summer drought-shrubs are evergreen with thick, waxy leaves to prevent water loss-fire is an important part of this biome-it may have a 40 year cycle-fire suppression by human interference has resulted in more accumulation of biomass so·the fires are often larger and more destructive when they occur.

6. Deserts-less than 10" rainfall/year-high temperatures increase evaporative water loss-least productive terrestrial ecosystem-sensitive to disturbance-plant reproduction may be a rare event-e.g., stand of mesquite in which all individuals are 250 years old-little or no reproduction in 250 years-desert plants and animals display numerous water conserving adaptations.

Aquatic Ecosystems: These can be classified into freshwater and marine ecosystems. Freshwater ecosystems include streams (lotic ecosystems), lakes and ponds (lentic ecosystems) and marshes and bogs (wetlands). Lotic ecosystems-moving water-speed of water current important for different aquatic species because it determines rate at which materials such as nutrients are flushed out of the system, it also determines the type of bottom (e.g., sandy, gravel, or rock), which, in turn, determines the kinds of plants and animals that can live there.

Lotic systems are largely open systems as water is constantly moving-nutrients such as N and P are limiting because they are flushed out-major input of carbon is in the form of leaf litter from nearby terrestrial ecosystems. Lentic systems-lakes and ponds-standing water-more closed than lotic systems and have lower rates of nutrients loss.

Lakes and ponds can also be divided into three life zones:

1. Littoral zone-closest to shore-contains rooted aquatic plants
2. Limnetic zone-sun lit body of lake

3. Profundal zone-below level of light penetration-no photosynthesis-decomposition is major process in this zone.

Lakes in temperate zones of the world display thermal stratification in the summer-surface layer (epilimnion) is heated by sun-temperature drops quickly in middle layer (thermocline)-bottom layer (hypolimnion)-these layers do not mix-epilimnion becomes nutrient poor as phytoplankton photosynthesise and use up nutrients-detritus (organic matter) sinks to bottom of lake-hypolimnion becomes oxygen-depleted as organic matter is decomposed.

Turnover occurs in fall as epilimnion cools, gets heavier and sinks—results in redistribution of oxygen and nutrients.

Lakes classified according to their nutrient status:

1. Oligotrophic-nutrient poor-low organic matter-low phytoplankton productivity.
2. Eutrophic-nutrient rich-high organic matter, high phytoplankton productivity-oligotrophic lakes can become eutrophic through inputs of nutrients from outside by human activity-e.g., sewage pollution, agricultural runoff.

Marine Ecosystems-begin at ocean's edge.

Estuaries-formed where large river flows into ocean-mixing zone of freshwater and saltwater-organisms must be adapted to changing salinity-they act as a nutrient trap-trap nutrients brought in by tides and carried down by river-because of their rich nutrients, estuaries have very high primary productivity-estuaries are often called "nurseries of the sea" because they are the place where many larval and immature marine fish grow up and develop before moving out into the ocean as adults.

Seashores-exposed to waves and tides-classified as rocky or sandy.

Rocky shore-offer something solid for plants and animals to attach to-seaweed (algae) are primary producers-animals include barnacles, oysters, sea urchins.

Sandy beach-most organisms dwell under the sand-e.g., ghost crabs, ghost shrimp, sandworms.

Coral reefs-diverse, productive ecosystems-found in shallow, tropical waters-reefs formed from calcium carbonate (limestone) exoskeletons of corals-crevices and caves provide shelter from many different plants and animals-sponges, echinoderms, fishes.

Ocean-covers roughly ¾ of earth's surface-can be divided into 3 life zones:

1. Neritic zone-shallow coastal waters-sunlight-seaweeds and phytoplankton abundant.
2. Pelagic zone-open sea-includes sunlit waters-water is nutrient poor and productivity is comparable to that of the terrestrial deserts-phyto-and zooplankton base of trophic structure-fed upon by fishes and whales.
3. Benthic-ocean floor-temperature about 2 degree Centigrade and pressures are intense-total darkness-bombarded by dead, organic matter sinking from above-instead of ecosystems based on photosynthesizing autotrophs, benthic ecosystems are mostly decomposers and organisms that feed on decomposers (detritivores).

Benthic habitats are also the location of a unique, recently discovered ecosystem.

The earth's surface consists of large tectonic plates-in areas of the ocean bottom where these plates are formed, heated water is released from cracks in the earth's crust-these areas are called hydrothermal vents-ecosystem associated with these vents are based not on photosynthetic plants (no light), but on chemosynthetic bacteria that are able to obtain energy from hydrogen sulfide (H_2S) released at the vents, these bacteria are the primary producers are fed upon by various consumers (worms and other invertebrates).

Gibbs Free Energy: The value of Gibbs Free Energy for a reaction is based upon its enthalpy and entropy. Both of these factors can have either a negative or positive effect on whether or not a reaction will occur spontaneously or at all. It can help to predict at what temperature a non-spontaneous reaction will become spontaneous. It is called free energy because it is the energy that will be released or freed up to do work. Gibbs Free energy is defined as:

$$G° = \Delta H° - T \Delta S°$$

It is also defined as G° = G-products-G-reactants. But this equation is only valuable if the reaction is carried out at standard temperature and pressure. Any other temperature requires the use of the first equation. A change can only be spontaneous if it is accompanied by a decrease in free energy. In other words, for a change to be spontaneous, G must be negative. When a change is exothermic and is also accompanied by an increase in entropy, both factors favour spontaneity., i.e., H is negative (-ve) S is positive (+ ve) $G = H - TS = (-) - T(+)$

In such a change, G will be negative regardless of the value of the absolute temperature, T (which can only have positive values). Therefore, the change will occur spontaneously at all temperatures. On the other hand, if a change is endothermic and is accompanied by a decrease in entropy, both factors work against spontaneity. i.e., H is positive (+ve) S is negative (-ve) $G = H - TS = (+)- T(-)$

In this case, G will be positive at all temperatures and the change will always be non-spontaneous. When H and S have the same sign, the temperature becomes critical in determining whether or not an event is spontaneous.

If H and S are both positive, $G = (+) - T(+)$

Only at relatively high-temperatures will the value of TS be larger than the value of H so that their difference, G, is negative. A familiar example is the melting of ice.

$$H_2O(s) \quad H_2O(1)$$

Here is a change that we know is endothermic and occurs with an increase in entropy. At temperatures above 0°C (when the pressure is 1 atm), ice melts because the TS term is bigger than the H term. At lower temperatures, ice doesn't melt because the smaller value of T gives a smaller value for TS and the difference H-TS, is positive.

For similar reasons, when H and S are both negative, G will be negative only at relatively low temperatures. The freezing of water is an example.

$$H_2O(1) \quad H_2O(s)$$

Energy is released as the solid is formed and the entropy decreases. Water freezes spontaneously at low temperatures, that is, below 0°C.

Potential of Redox

To understand it one should know the basics of redox potentiometry.

Fundamental of Redox Potentiometry: Oxidation reduction (or redox) reactions involve the transfer of an electron (or H, or hydride group) from a donor to an acceptor. The donor loses an electron, and changes from the reduced to the oxidised form; the acceptor gains an electron, and changes from the oxidised to the reduced form. The oxidised and reduced form of a redox component are said to form a redox couple.

The special interest of redox reactions in thermodynamics lies in the fact that the free energy change can be measured directly using an electrochemical cell.

Overall reaction:

$$A^{-}+B \Leftrightarrow A+B$$

In an electrochemical cell the components of the reaction are separated so that the donor couple and acceptor couple are in separate reaction vessels. A salt bridge connects both vessels, so that the electrostatic potential difference between them is zero. In each vessel, an electrode (usually made of Pt) connects the reaction mixture in one cell to that in the other, through the measuring device (a potentiometer or electrometer).

The electrochemical cell therefore consists of two half-cells, each of which contains the reactants (oxidised and reduced components) of one redox couple, making up the half-cell reactants. In principle, the circuit allows the electrons to flow from one half-cell to the other, down the gradient provided by the difference in free energy of the electrons (measured in volts) in equilibrium with the two half-cell reactions. By channeling this flow through the measuring device, the system is constrained so that the flux is zero (or very nearly so).

If a potentiometer is used, the condition of zero flux is obtained by balancing the voltage difference between the two half-cell reactions through an applied potential of opposite sign, adjusting the current, measured with a sensitive galvanometer (or other suitable device), to zero. If an electrometer is used, the voltage difference between the half cell reactions is measured directly, the condition of zero flux being achieved by use of an electrometer of very high input impedance. In either case, the voltage measured shows the driving force represented by the difference in free energy of the electrons in equilibrium with the two half-cell reactions. This difference is called the electrode potential difference, the oxidation reduction potential difference, or the redox potential difference.

Because the redox potential difference is measured under conditions of zero flux (the "reversible" conditions of classical thermodynamics), it is a direct measure of the free energy change of the reaction of the electrochemical cell. The electrode potential measures the intensity factor (DE, measured in volts) of the work term (DG′ = -zFDE) for the process represented by the oxidation reduction reaction of the electrochemical cell (see below). In the case of the potentiometric measurement, this can be seen clearly because the voltage supplied by the potentiometer balances the electrode potential difference by application of external work from a change in state in the surroundings (the potentiometer).

The Reference Electrode: In practice, the redox potential difference is measured with respect to a standard electrode. Several sorts of electrodes are commonly used. The standard electrode is a hydrogen half-cell with the reaction

$$H_2(\text{gas}) \Leftrightarrow 2H^+ + 2e^-$$

in which all components are in their standard states (1 atm pressure for the gas, 1M activity for the proton, or pH 0). This is not a convenient device to handle, so more often a secondary standard reference electrode is′ used, a calomel electrode or a Ag/AgCl electrode. These latter have redox potentials with respect to the standard H_2-electrode which are well characterised. The redox potential for a couple is usually assumed to be the value with respect to the standard reference electrode (the hydrogen electrode). If a value is measured with respect to a secondary standard, the secondary reference electrode potential with respect to the standard is added to this value. Free energy change for an oxidation reduction reaction We will consider a redox reaction

$$A^- + B \Leftrightarrow A + B^-$$

where the two half-cell reactions are those of the diagram above. We will assume that the B/B^- couple is a standard reference electrode.

Let us consider the half-cell reaction for the couple A/A^-,

$$A^- \Leftrightarrow A + e^-$$

and suppose that at activities [A] and [A^-], an electrode potential of E volts is measured with respect to the reference half-cell. If the concentrations are changed to [A] and [A^-], a different electrode potential of E′ volts is measured. Under the conditions of measurement (zero flux), the reaction of eq. 1 is at equilibrium, in each case, with the free energy of the electron measured by the electrode potential. Chemical free energy of A^- = chemical free energy of A + electrical free energy of e^- or more concisely (and rearranging terms)

$$\mu_A - \mu_{A-} - zFE = 0 = \mu'A - \mu'A- - zFE'$$

Here z is the number of electrons involved in the reaction (in this case 1), and F is the Faraday constant. (The sign of E is negative because the electron carries a negative charge.)

Relation between Electrode Potential and Free Energy of Reaction: The complete reaction of an electrochemical cell can be treated like any other reaction, using the equation for DG′. Since the activity terms in the logarithmic ratio are the same as the activity terms in the equations for E above, it is a straightforward exercise to substitute between equations to find the relation between E and DG. This gives the following relationships:

$$DG' = - zFDE'$$

$$DG^\circ = - zFDE^\circ$$

$$DG^{\circ\prime} = - zFDE^{\circ\prime}$$

Assessment of Gravimeter

Gravimetric analysis a method of quantitative chemical analysis in which the constituent sought is converted into a substance (of known composition) that can be separated from the sample and weighed. The steps commonly followed in gravimetric analysis are: (1) preparation of a solution containing a known weight of the sample, (2) separation of the desired constituent, (3) weighing the isolated constituent, and

(4) computation of the amount of the particular constituent in the sample from the observed weight of the isolated substance.

Of the many methods that may be employed for isolating the desired constituent from a solution of a sample, the most common is precipitation—that is, transformation into a substance not soluble in the solution. A reagent is added that forms an insoluble compound with the desired constituent but will not precipitate other constituents of the sample. The precipitate obtained is separated by filtration, washed free of soluble impurities, dried or ignited to remove water, and weighed.

Certain substances can be separated by virtue of their easy convertibility into gaseous compounds, as in the determination of carbonate in a mineral analysis. The sample is treated with an acid, and carbon dioxide is evolved as a gas. The gas is absorbed on a weighed quantity of a solid alkaline reagent, and the amount of carbon dioxide is determined from the gain in weight of the absorbent. Electrodeposition is used in order to separate certain metals that can be plated out by passing an electric current through a solution of their salts. Copper in alloys may be determined by this method as long as the sample is free from other metals that plate out under the same conditions. Errors made in gravimetric analyses usually relate to the purity of the isolated constituent. In general, the compounds that are precipitated are very insoluble, and negligible error results from the incompleteness of precipitation. Obtaining a precipitate that is 100 per cent pure and is exactly of the composition represented by a chemical formula is, however, considerably more difficult. All gravimetric methods are subject to some degree of error because of this difficulty.

Titrimetric Analysis

Types of Titrimetric Analysis: a. Acid-base Titrations; b. Redox Titrations.

Titrations Based on Acid

Acid can be defined as a reagent which donates protons (H+) and a base as one which accepts protons. An acid-base reaction therefore involves the transfer of one or more protons. Consider the following reaction:

$$HCl + NaOH = NaCl + H_2O$$

The important thing here is that one "molecule" of HCl reacts with one molecule of NaOH. (Actually both are ionised in solution but for our purposes the above description is adequate). In terms of a titration, we might have the HCl in the conical flask (the titrand) and NaOH in the burette (the titrant).

Example: Suppose that the NaOH was in the burette and that this had been earlier standardised (i.e. its exact concentration is known) and is 0.0985 mol dm^{-3}. There are 25.0 cm^3 of HCl in the conical flask (delivered by pipette). The equivalence point was reached when 27.25 cm^{-3} of NaOH was added, what is the concentration of HCl ?

First Stage: How many moles of NaOH were required to reach equivalence point?

The concentration of the NaOH is 0.0985 mole dm-3 and 27.25 cm^{-3} were added so the number of moles added was:

$$= (0.0985 \times 27.25)/1000 = 2.684 - 10^{-3}$$

Since one molecule NaOH reacts with one molecule HCl (likewise one mole and one mole), there must be the same number of moles of HCl as there are NaOH = 2.684 x 10^{-3}.

OK, this is the number of moles of HCl is the conical flask, i.e. in 25.0 cm^{-3}. Usually you are asked for concentration in terms of mole dm ^{3} so this figure must be multiplied by 1000/25. (Concentration volume required/concentration volume actual). So the concentration (mol dm^{-3}) is ?

Exercises

1. A solution of HCl (25.0 cm^3) was titrated against NaOH (0.102 mole dm^{-3}) given the titre was 25.40 cm^3, calculate the concentration of the HCl solution.

   (**Answer:** 0.104 mol dm^{-3})

2. In another titration using the NaOH from above (0.102 mole dm^{-3}) against sulphuric acid (25.0 cm^3) the titre was 28.60 cm^3, calculate the concentration of the sulphuric acid.

   (**Answer:** 0.058 mole dm^{-3})

3. Arsenic (V) acid (H_3ASO_4), is a tribasic acid, 25.0 cm^3 of a solution of this acid required 35.7 cm^3 of sodium hydroxide (0.100 mol dm^{-3}) for titration. Calculate the concentration of H_3ASO_4.

   (**Answer:** 0.0476 mol dm^{-3} or 4.76 x 10^{-2} mol dm^{-3}).

4. If a monoacidic base of concentration 0.1000 mol dm^{-3} is titrated against 25 cm3 of a dibasic acid (0.0460 mol dm^{-3}) what volume of base will be required to reach equivalence point ?

   (**Answer:** 23.0 cm^3)

Redox Titrations and their Part: "Redox" stands for reduction-oxidation. In this type of reaction one species is reduced (that is a reduction in its oxidation number-or a gain of electrons for example M^3+ going to M^2+ whilst the other is oxidised (an increase in its oxidation number or a loss of electrons for example M^2+ going to M^3+). Oxidation number is defined as the charge which an atom appears to have (assuming the bonding to be entirely ionic) when electrons are counted using the following rules.

Electrons shared between like atoms are equally divided between them. Hence all elements e.g., H_2, O_2, etc. have an oxidation number (or state) of zero.

Electrons shared between unlike atoms are counted with the more electronegative atom e.g., in HCl the electrons are assumed to be associated with the chlorine so the oxidation numbers are hydrogen +1, chlorine -1.

To make it easier use the following rules; The alkali metals (Group I) always are +1 Group II metals are +2.

Oxygen is -2 (except in peroxides)

Hydrogen is +1 (except with Group I and Group II when it is -1)

Example: ON of Mn in $KMnO_4$.

K = +1; 4 × O = (4 x -2) = - 8

So for electrical neutrality the ON on Mn must be +7.

Balancing Redox Equations

Oxidation involves an increase in ON and a loss of electron(s) Reduction involves a decrease in ON and a gain of electron(s)

Oxidising agents cause a decrease in ON and the agent gains e^{-1} s. Reducing Agents cause an increase in ON and the agent loses e^{-1}s. There are a number of methods used in balancing redox reactions.

Oxidation Number Method

1. Assign ON's to substances for which ON changes.
2. Decide on the number of electrons to be moved per atom.
3. Decide on the number of electrons to be moved per formula unit.
4. Since electrons gained by the oxidisable reactant must be supplied by the reducible reactant, compensate electron gain or loss by inserting appropriate coefficients to the oxidising and reducing agents. In

other words, get the ratio of the reacting species correct in terms of the electrons gained or lost.

5. Insert other coefficients to be consistent with the conservation of matter. Let's take a look at an example:

The reaction between $K_2Cr_2O_7$ and H_2SO_3 in acid solution to give $KHSO_4$ The basic "equation" might be written as:

$$K_2Cr_2O_7 + H_2SO_3 + H+ = KHSO_4 + Cr^3+ H_2O$$

For Cr:

1. For the reactant ($K_2Cr_2O_7$) the oxidation number on Cr = +6 For the product (Cr^{3+}; the oxidation number on Cr = +3
2. So we need 3 electrons (+6 goes to +3) for each Cr atom.
3. Since there are two Cr per formula unit we need 6 electrons per formula unit.

For S:

1. In H_2SO_3 ON for S = +4
   In $KHSO_3$ ON for S = +6
2. So for each S atom undergoing change there are 2 electrons released therefore 2 electrons per formula unit.
3. We now know that $K_2Cr_2O_7$ and H_2SO_3 must react in the ratio 2:6 or 1:3 We can now write;

$$K_2Cr_2O_7 + 3\,H_2SO_3 + H^+ = 3\,HSO_4^- + 2\,Cr^{3+} + 2K^+ + H_2O$$

We now need to balance H and O.

On the left hand side we have 16 O and on the right we have 12 O, so we need $4H_2O$ to make up the difference. This leaves an imbalance of H+ which can be made up simply be adding H+ to the lefthand side.

$$K_2Cr_2O_7 + 3\,H_2SO_3 + 5\,H^+ = 3\,HSO_4 + 2\,Cr^{3+} + 2K^+ + 4\,H_2O$$

Examples of some important Redox Titations:

Peroxydisulphate Ion

The peroxydisulphate ion (or persulphate ion) $S_2O_8^{-2}$ (usually the potassium or ammonium salt) is a strong oxidising agent in acid medium:

$S_2O_8^{2-} + 2e^- > 25O_4^{2-}$ E = +2.0V

Any excess of this oxidant may be removed by boiling;

$5_2O_8^{2-} + 2H_2O > SO_4^{2-} + 2H^+ + H_2O_2\ 2H_2O_2 > H_2O + O_2$

The important uses of this oxidant include the conversion of manganese to permanganate, chromium to dichromate and vanadium to vanadate.

Perchloric Acid

Hot concentrated perchloric acid is a strong oxidant;

$2ClO_4^- + 16H+ +14e^- > C1_2 + SH_2OE = +2V$

Used to oxidise chromium and vanadium ores/alloys to dichromate and vanadate. Permanganate as an oxidising agent. The permanganate (or manganate (VII)) ion is one of the most common oxidising titrant. In strongly acidic media, permanganate undergoes a five electron reduction.

$MnO_4^- + 8H^+ + 5e^- > Mn^{2+} + 4H_2O$ E = +1.5V

In weak acid, neutral or alkaline conditions permanganate undergoes a three electron transition to manganese (IV)

$MnO_4^- + 4H^+ + 3e^- > MnO_2 + 2H_2O$ E = +1.69V

Or in alkaline solution;

$MnO_4 + 2H_2O + 3e^- > MnO_2 + 4OH^-$ E = + 0.57V

Note: Permanganate solutions are unstable, therefore not a primary standard.

$4MnO_4^- + 2H_2O > 4MnO_2 + 4OH^- + 3O_2$ *(MnO_2 catalyses this* *reaction)* Permanganate must be standardised for example with oxalate;

$H_2C_2O_4 > 2H^+ + CO_2 + 2e$

Overall: $2MnO_4 + 5H_2C_2O_4 + 16H^+ > 2Mn^{2+} + 10CO_2 + 8H_2O$

Initially the reaction is slow but is catalysed by Mn ²+ so becomes more rapid. Permanganate titrations:

Hydrogen peroxide in acidic conditions;

$5H_2O_2 + 2MnO_4 + 6H^+ > 2Mn^{2+} + 5O_2 + 8H_2O$

Iron (II) in acidic conditions is easily oxidised by permanganate

$MnO_4^- + 5Fe^{2+} + 8H^+ > Mn^{2+} + 5Fe^{3+} + 4H^2O$

Dichromate Reactions: Dichromate ion is an oxidising agent

$Cr_2O_7^{2-} + 14H^+ + 6e > 2Cr^{3+} + 7H_2OE = +1.33V$

It can be prepared as a standard solution and so avoids the need to standardise as is the case with permanganate.

Iodine Methods

$I_2 + 2e^- > 21^- E = +0.54V$

Value for E is intermediate can therefore be reduced or oxidised. Iodine can be reduced to iodide by for example As (III), Sn (II) whilst iodide can be oxidised to iodine by for example permanganate.

Thiosulphate

In acid, neutral or slightly alkaline solutions;

$25_2O_3^{2-} > S_4O_6^{2} + 2e^-$

$2S_2O_3^{2-} + I_2 > S_4O_6^{2-} + 2I^-$

strongly alkaline solutions;

$2S_2O_3^{2-} + 4I_2 + 10OH^- > 2S_4O_6^{2-} + 8I^- + 5H_2O$

Iodate

$IO_3^- + 5I^- + 6H^+ > 3I_2 + 3H_2O$

Hydrogen Peroxide: Strong oxidant even in alkaline conditions.

$H_2O_2 + 2H^+ + 2e^- > 2H_2O \; E = +1.77V$

Danger of Chemical Carcinogenesis: In 1775, an English physician named Percival Pott observed that the high incidence of cancer of the scrotum in chimney sweepers was due to their exposure to coal tars. A century later, scientists in Germany found a high incidence of bladder cancer among workers exposed to aromatic amines (chemicals used in the dyeing and pigment industry).

Since then, researchers have identified many more substances that are known or suspected to cause cancer. These substances are referred to as carcinogens. Presently in 25-30,000 chemicals are manufactured at commercial level or generated as waste and emissions released during industrial processes. Now pollution of human environment is suspected as a major cause of cancer.

Factors Carrying Risk of Cancer: Cancer is often perceived as a random, and unpredictable disease. In fact, it's estimated that about 80 per cent of all cancers are caused by "behavioral" factors and are therefore under human control. Two of the leading risk factors associated with cancer are tobacco smoke and diet. Together they are responsible for about two thirds of all new cancer cases. Tobacco smoke is the major cause of 80 to 90 per cent of lung cancer cases. It is also an important cause of bladder and breast cancer. Diet associated cancers include colon, rectum, breast and stomach cancer. For example, high fat consumption is considered a risk factor for colon and possibly breast cancer. Other well-established risk factors include work-related exposure, alcohol consumption, radiation including sunlight, drugs and viruses.

By comparison, the role of chemical pollutants is unclear. Scientists have estimated that chemicals in the environment account for about two per cent of all cancer deaths. There is little direct evidence linking exposure to specific pollutants in the general environment and the incidence of cancer.

Determining the Cancer Risk or Carcinogenicity: To determine the potential of chemical pollutants to cause cancer, scientists rely primarily on laboratory studies in which animals are exposed to chemicals, and when available, population studies involving human exposure to chemicals.

Chemicals can be grouped into six categories: "carcinogenic to humans", "probably carcinogenic to humans", "possibly carcinogenic to humans", "unlikely to be carcinogenic to humans", "probably not carcinogenic to humans" and "unclassifiable" with respect to cancer causing ability. Scientists assess carcinogenicity by weighing the available data from both animal and human studies.

So far, out of thousands of chemicals tested, scientists have identified only a few hundred that cause cancer in rodents and around 50 that definitely cause cancer in humans. The number of confirmed human carcinogens is lower because, for the majority of chemicals, there is often little or no information available on cancer incidence rates in humans. Moreover, effects that occur in laboratory animals will not necessarily occur in humans. In other words, what is harmful to a rat or mouse may not be harmful to a human.

Atmospheric Carcanogens

In theory, there is no "safe" level of exposure to a carcinogen. It is generally assumed, but has never been proven, that there is some chance of developing cancer at any level of exposure. In other words, contact between a single human cell and a carcinogen molecule may be enough to cause cancer. This does not mean that cancer is the inevitable or even the likely outcome of such an event. Human cells have developed a range of defenses to guard against all sorts of injuries, including those that lead to cancer. As a result, cancer usually occurs only after prolonged and continuous exposure to carcinogens. In addition, as the exposure levels decrease, so does the risk of developing cancer.

People today are exposed to trace amounts of various chemicals in their air, water and food. But the levels are generally far below those levels associated with harmful effects. In the workplace, however, exposure to much higher levels of chemicals can occur.

In 1989, the government of Canada listed a group of 44 commercial chemicals, known as priority substances list, of the 44 substances on the List, 19 were considered "toxic" to human health. These substances were considered toxic largely based on their potential to cause cancer. The actual risk of a chemical causing cancer depends on two factors: its ability to cause cancer, and the levels to which people are exposed.

A partial list of substances known to be toxic or carcinogenic to human health is given below:

- Arsenic (Inorganic Arsenic)
- Benzene
- Benzidine
- Bis (2-Chloromethyl) Ether
- Cadmium (Inorganic Cadmium)
- Chlorinated Paraffin Waxes
- Chloromethyl Methyl Ether
- Chromium (Hexavalent Chromium)
- 3, 3-Dichlorobenzidine
- 1, 2-Dichloroethane
- Bis (2-Ethylhexyl) Phthalate (DEHP)

- Dichloromethane
- Hexachlorobenzene
- Mineral Fibres (Refractory Ceramic Fibre)
- Nickel ("oxidic", "sulphide", and "soluble" forms)
- Polycyclic Aromatic Hydrocarbons
- Polychlorinated Dibenzodioxins
- Polychlorinated Dibenzofurans
- Trichloroethylene

Elements in Atmosphere

The constituents of atmosphere can be divided into three categories-major, minor and trace. The composition of pollution-free dry air at ground level on volume basis is as follows:

Major components	Nitrogen	78.09%
	Oxygen	20.09%
	Water Vapor	0.1- 5% in wet air
Minor Components	Argon	0.934%
	Carbon dioxide	0.0350% or 350 *ppm*
Trace Components	Neon	18 *ppm*
Helium	5.24 *ppm*	
Methane	2 *ppm*	
Krypton	1.14 *ppm*	
Nitrous Oxide	0.25 *ppm*	
Hydrogen	0.5 *ppm*	
Xenon	0.08 *ppm*	
Sulphur dioxide	0.0002 *ppm*	
Nitrogen dioxide	0.1 *ppm*	
Carbon monoxide	0.12 *ppm*	
Ammonia	0.01 *ppm*	
Ozone	trace	

With altitude the composition of atmosphere changes. Density and pressure both decrease with increase in altitude. The pressure decreases from one atmosphere at sea level to 3×10^{-7} atmosphere at 100 km above sea level.

Making of Atmosphere

The atmosphere can be divided into four regions on the basis of temperature range and chemical species.

Table Regions of the Atmosphere

Region	*Altitude range, km*	*Temperature range, °C*	*Important Chemical species*
Troposphere	0-11	15 to - 56	N_2, O_2, H_2O, CO_2
Stratosphere	11-50	- 56 to - 2	O_3
Mesosphere	50-85	- 2 to - 92	O_2+, NO^+
Thermosphere	85-500	- 92 to 1200	O_2+, O+, NO

(a) *Trophosphere:* Trophosphere is the zone of life. It contains 70 per cent of the mass of atmosphere. The composition of trophosphere is almost homogeneous, in absence of air pollution, due to the constant motion of air masses in this region. Trophosphere has a negative lapse rate, i.e., temperature decreases with increase in altitude.

(b) *Stratosphere:* Stratosphere has a positive lapse rate, i.e., the temperature increases with increase in altitude. The positive lapse rate in this zone is caused by absorption of ultraviolet radiation by ozone present in this zone.

O_3 + UV radiation $\rightarrow O_2 + O°$

As UV radiation is injurious for the living organisms on the earth's surface, stratosphere acts as a protective shield for life on earth by absorbing UV radiation.

Movement inside stratosphere is very slow, so the residence times of molecules or particles in this region are long. Due to high residence times pollutants reaching in this region pose long-term global hazards.

(c) *Mesosphere:* It shows negative lapse rate, which is due to low levels of ultraviolet absorbing ozone and other species.

(d) *Thermosphere:* Thermosphere has a positive lapse rate. The temperature increases upto 1200°C. In this zone oxygen and nitric oxide absorb for ultraviolet portion of solar radiation and split into atoms or ionise.

Particles in the Atmosphere: Particles in the atmosphere range in size from about one half of a millimeter down to molecular dimensions, examples are combustion nuclei, sea salt nuclei, cement dust, soil dust, pulverised coal, bacteria, virus, fungal spores, pollen, etc.

Some important terms regarding atmospheric particles:

Term	*Meaning*
Aerosol	Colloidal-sized atmospheric panicle
Condensation aerosol	Formed by condensation of vapors or reactions of gases
Dispersion aerosol	Formed by grinding of solids, atomisation of liquids or dispensation of dusts
Fog	Term denoting high level of water droplets
Haze	Denotes decreased visibility due to the presence of particles Mists Liquid particles
Smoke	Particles formed by incomplete combustion of fuel.

Physical Processes for Particle Formation: Dispersion aerosols such as dusts are formed from the disintegration of larger particles and are usually above 1 mm in size. Ex. Blowing of dirt from dry soil, dust formation during coal and stone grinding, etc. Many dispersion aerosols originate from natural sources such as sea spray, wind blown dust and volcanic dust. Most dispersion aerosols are relatively large so are less harmful than smaller ones.

Formation of Particles through Chemical Process: Chemical processes in the atmosphere form large quantities of particulate matter. Organic pollutants and nitrogen oxides are mostly involved in these reactions. Most chemical processes that produce particles are combustion processes like fossil fuel based power plants, incinerator, cement kilns, I.C. engines, forest fires and active volcanoes. Mostly the size of these particles is below 1 mm, so they can be easily carried into the alveoli of lungs.

Particles of Inorganic Nature

(i) Metal oxides are formed whenever fuels containing metals are burnt. Ex particulate iron oxide is formed during the combustion of pyrite-containing coal.
$3 FeS_2 + 8O_2 \rightarrow Fe_3O_4 + 6SO_2$

(ii) During coal combustion a part of the calcium carbonate is converted to calcium oxide and is emitted in the atmosphere.
$CaCO_3 + heat \rightarrow CaO + O_2$

(iii) Organic vanadium in residual fuel oil is converted to particulate vanadium oxide during combustion.

(iv) Very commonly aerosol mists form due to oxidation of atmospheric sulphur dioxide to sulfuric acid which accumulates atmospheric water to form small liquid droplets.

$2 SO_2 + O_2 + 2 H_2O \rightarrow 2 H_2 SO_4$

This sulfuric acid reacts with basic air pollutants such as ammonia or calcium oxide to form salts.

$$H_2SO_4 \text{ (droplet)} + 2NH_3 \text{ (g)} \rightarrow (NH_4)_2 SO_4 \text{ (droplet)}$$

$$H_2SO_4 \text{ (droplet)} + C\alpha O \text{ (g)} \rightarrow CaSO_4 \text{ (droplet)} + H_2O$$

Under dry conditions water is lost from these droplets and a solid aerosol is formed.

Particles of Organic Nature: A significant fraction of organic particulate matter is produced by internal combustion engines. These products may include nitrogen-containing compounds and oxidised hydrocarbon polymers. Gasoline autoengines and diesel truck engines emit more than 100 compounds. The prominent classes are -n-alkanes, n-alkanoic acids, benzaldehydes, benzoic acids, azanaphthalenes, polycyclic aromatic hydrocarbons, oxygenated PAHs, etc.

PAH Synthesis: PAH hydrocarbons are of greatest concern. They consist of condensed ring aromatic molecules (aryl). The most common PAH compound is benzoapyrene, which is carcinogenic.

PAHs may be synthesised from saturated hydrocarbons under oxygen-deficient conditions. Hydrocarbons with very low molecular masses, including methane may act as percursors for the PANS by pyrosynthesis.

At temperatures greater than 500 °C C - H and C - C bonds are broken to form free radicals. These radicles undergo dehydrogenation and combine to form aryl ring structures which are resistant to thermal degradation. The tending of hydrocarbons to form PAHs by pyrosynthesis varies in the order

Aromatics > Cycloolefins > Olefins > paraffins.

PAHs may be formed from higher alkanes present in fuels and plants by the process of pyrolysis or crackiting.

Important Functions of Particulates in the Atmosphere: Particles in the size range 0.1 - 1 μ exert some important effects given below.

(a) They serve as nuclei for the formation of clouds, fog, water droplets and ice crystals.

(b) They play an important role in maintaining the heat balance of the earth's atmosphere by reflection of solar radiation.

(c) They are involved in several chemical reactions in the atmosphere for example:

(i) Neutralisation reactions in the water droplets

(ii) Catalytic effects of metal oxide particulates on oxidation reactions

(iii) Photochemical oxidation reactions. Ions and radicals in the atmosphere.

Above stratosphere ions are so prevalent that mesosphere and thermosphere are collectively called as Ionosphere. Both positive ions (0_2^+, O^+, NO^+, etc.) and electrons exist at significant levels in the ionosphere. Due to rarefied conditions in the upper atmosphere these ions persist for long periods without recombining to produce neutral species.

Ultraviolet radiation is primarily responsible for the production of ions in the ionosphere. In darkness positive ions slowly recombine with free electrons to form the original neutral species. These ions are strongly influenced by the earth's magnetic field, and form Van Allen belts. Van Allen belts are two belts of ionised particles encircling the earth. The inner belt consist of protons and the outer belt consist of electrons. Ions can also be produced in troposphere by weather phenomena. Besides ions, energetic electromagnetic radiation in the atmosphere may produce atoms or groups of atoms with unpaired electrons called free radicals.

$$\begin{matrix} O \\ \| \\ H_3C\text{-}C\text{-}H + hv \end{matrix} \rightarrow H_0C^{\bullet} + {}^{\bullet}\begin{matrix} O \\ \| \\ C\text{- }H \end{matrix}$$

Because of their unpaired electrons they are highly reactive. They take part in chain reactions in which one of the products of each reaction is a free radical. When the free radical is destroyed by reaction with another free on radical, the chain ends.

$$H_3\,C. + H_3\,C. \rightarrow C_2\,H_6$$

Reaction of Photochemicals: The absorption of high energy solar radiation like ultraviolet radiation by chemical species can bring about reactions, thus known as photochemical reactions, which otherwise do not occur in normal conditions in absence of light. Photochemical reactions play an important role in determining the nature and fate of chemical species present in the atmosphere.

When molecules absorb energetic electromagnetic radiation in the UV or visible region of solar spectrum electronically excited molecules are produced.

Due to high energy, electrons of the molecule are promoted to a vacant orbital of higher energy, which makes them highly reactive. The reaction of electronically excited species are determined by the way in which they loose their energy. For example

(a) Loss of energy to another molecule or atom

$O_2^* + M \rightarrow O_2 + M$ (high energy)

(b) Dissociation of the excited molecule

$O_2^* \rightarrow O + O$

(c) Ionisation by loss of an electron

$N_2^{\bullet} \rightarrow \dot{N}_2^{\;+} + e^-$

(d) Direct reaction with other species

$O_2^* + O_3 \rightarrow 2O_2 + O$

Photochemical reactions in the atmosphere may lead to photochemical smog formation.

Oxygen and Ozone Chemistry: Due to its chemical activity oxygen is very important in trophosphere. Oxides are the stable forms of most elements. Oxygen cycle has a key role in life processes, atmospheric chemistry and geochemical reactions. It is based on the exchange of oxygen between atmosphere, lithosphere, hydrosphere and biosphere.

Oxygen is used in combustion, oxidative respiration, and oxidative weathering processes.

C (in organic matter) $+ O_2 \quad O_2 \xrightarrow{\text{Combustion}} CO_2$

$\{CH_2O\} + O_2 \xrightarrow{\text{repiration}} CO_2 + H_2O$

$4\,FeO + O_2 \rightarrow 2\,Fe_2O_3$

Plant return oxygen to atmosphere by photosynthesis

$CO_2 + H_2O + \text{Sunlight} \rightarrow \{CH_2O\} + O_2$

Due to rarefied atmosphere and ionizing radiations stratosphere contains Oxygen atoms, O; excited oxygen atoms O*, oxygen ions O+, excited oxygen molecule, O_2^*, ozone, O_3; and oxygen, O_2. These chemical species are formed by following photochemical reactions

$O_2 + hv \xrightarrow{\text{135–176 nm wavelength or}} O + O$

$O_3 + hv \xrightarrow{\text{less than 308 nm wavelength}} O^* + O_2$

$O + hv \rightarrow O^+ + e^-$

$O_2 + O^+ \rightarrow O_2^+ + O$

$O^+ + N_2 \rightarrow NO^+ + N$

$O_2 + hv \rightarrow O_2^+ + e^-$

Ozone is formed by the following three body reactions.

$O + O_2 + M (N_2 \text{ or } O_2) \rightarrow O_3 + M$ (increased energy)

The maximum concentration of ozone is 10 ppm at 25-30 km height in stratosphere. Ozone absorbs the W radiation and forms a protective shield for the life on the earth. Ozone is an unstable gas and breaks back to molecular O_2

$$2O_3 \rightarrow 3O_2$$

Breakdown of ozone is catalyzed by NO, NO_2, H, HO°, HOO°, CIO, Cl, Br and BrO. Stratospheric ozone reacts with atomic oxygen, hydroxyl radical and NO.

Remote Sensing: Remote sensing is the science of acquiring information about an object or a surface feature without coming in physical contact with the object or feature of interest.

Remote sensing technology involve electromagnetic energy, acoustic waves and force fields to gain information about an object.

Force Fields: Gravity and magnetism are frequently used by geologists, geophysicists and astronomers for various purposes

Acoustical Waves: Very often acoustical waves are used for examining the ocean floor. An acoustic ecosounder releases short pulses of audible sound in a selected direction and the returning signals are collected to gain information about sea floor.

Electromagnetic Energy: It is the most commonly used medium for remote sensing.

Applications of Remote Sensing in Environmental Sciences: From environmental point of view, remote sensing is used in following areas:

(a) Weather analysis and forecasting

(b) Global climatological studies

(c) Hydrological studies

(d) Soils and landforms studies

(e) Rocks and mineral resources studies

(f) Ecology, conservation and resource management

(g) Land use and crop production

(h) Built environment studies

(i) Hazards and disasters prediction and monitoring:

(a) *Analysis and Forecasting of Weather:* TIROS 1 (Television and Infrared Observation Satellite) launched in 1960 was the first weather satellite. Since then more than 100 satellites have been launched for weather monitoring. The world meteorological organisation runs a "world weather watch" program based on observations from such satellites.

(b) *Studies on World Climatology:* The phenomena of "global warming" and "ozone hole" had increased emphasis on global climatological studies. Due to their monitoring on global scale along with other features they are indispensable tool in global climatological studies. They are used to find earth's energy and radiation budget, global atmospheric moisture distributions, wind flows and air circulations over the earth.

(c) *Studies on Hydrology:* Water is the basic requirement of living organisms and very important for all countries of the world. Hydrological studies can be divided into following four areas.

(i) *Hydrometeorology:* It is concerned mostly with the exchange processes in hydrological cycles. Microwave radars have been used since 1940 for rainfall monitoring. Ice and snow are monitored by satellites. Satellite data like solar radiation, surface albedo, surface soil moisture, etc. are used to estimate evaporation and evapotranspiration.

(ii) Surface hydrology focuses on water at or near the surface of the land. Landsat and SPOT image analysis are used for deciding watershed parameters. Meteostat visible and infrared imagery have been used in monitoring water bodies. Major floods throughout the world are monitored by satellites. Remote sensing had helped in identifying pollution sources in coastal zones and the water quality evaluation.

(iii) Hydrogeology is concerned with water at or below the surface of rock and regolith. Remote sensing is of value in large-scale preliminary investigation of groundwater reserves on the basis of soil moisture, vegetation, geological structures, etc. India has used landsat MSS data to decide promising locations of groundwater reserves.

(iv) Oceanography deals with the nature and behaviour of oceans. NOAA and Meteosat and other satellites have been used to monitor oceans for sea-surface temperature (SST) and circulation patterns. Water quality and salinity are assessed using satellite data.

(d) *Studies on Soils and Landforms:* Remote sensing can record surface colour, surface roughness, soil temperature and soil moisture. Aerial photography is commonly used for soil mapping. Data from landsat satellites have been used extensively in terrain studies. Visible and near infrared scanner data were used for such studies.

(e) *Rocks and Minerals:* Using special satellite imageries the rock formations and the available minerals present in particular area can be easily studied.

(f) *Studies on Ecological and Resource Management:* Satellites are finding uses in keeping surveillance of reserved areas. Onset of diseases, illegal felling can be easily detected.

(g) *Use of Land and Crop Production:* Satellite imageries are fastest ways to study the land use pattern. Crop yield forecasting uses satellite information.

(h) *Built-environment Studies:* Aerial photographs and satellite imageries are being used in public utility projects like water distribution, and sewage line laying. Aerial photographs of cities are used in road planning and the traffic studies. Air photography of industrial areas is frequently recorded at scales larger than 1 : 10,000. Different remote sensing techniques have been developed to measure the concentration and spread of pollutants e.g.,

  (i) *Visible Waveband Photography:* It is used to study the behaviour of chimney plumes and forest fires.

  (ii) *Infrared Imagery:* It is used for monitoring forest fires and spread of warm effluents in rivers, lakes or the sea.

  (iii) *Microwave Technique:* It is used to map oil spills in seas.

  (iv) *Multispectral Processing Technique:* It is used to assess water quality.

Remote sensing can provide valuable information concerning population levels and changes in it. Urban areas can be classified and mapped using the high spatial resolution data from landsat TM and SPOT multispectral or panchromatic systems. Land use changes can also be monitored by remote sensing.

Geographic Information System (GIS): The function of an information system is to improve one's ability to make decisions. An information system is that chain of operations that takes us from planning the observation and collection of data, to storage and analysis of data, to the use of the derived information in some decision-making process. A geographic information system (GIS) is an information system that is designed to work with data referenced by spatial or geographic coordinates.

A geographic information system can be either manual or automated (based on a digital computer). Manual GIS usually comprises several data elements like maps, aerial photographs, statistical reports, field survey reports, etc. These set of data are compiled and analysed with instruments like stereo-viewers, planimeters, etc.

A GIS must Contain following Five Elements

(a) *Data Acquisition:* It is the process of identifying and gathering the data required for the specific application. For this surveys of specific kinds, maps, aerial and ground photography, etc. are used.

(b) *Preprocessing:* It involves manipulating the data in several ways so that it may be entered into the GIS. The original data available in the form of maps, photographs and printed records is recorded in a computer database.

(c) *Data Management:* It governs the creation of, and access to, the database itself. These functions provide consistent methods for data entry, update, deletion and retrieval.

(d) *Manipulation and Analysis:* This portion of a GIS contains analytic operators that work with the database contents to derive new information.

(e) *Product Generation:* It is the phase where final outputs from the GIS are created, The output products may be in the form of statistical tables (e.g., a table showing population density in each district of India), maps (e.g., a presentation showing the most suitable place in Uttar Pradesh where Rhinoceros can be reintroduced), and graphics of various kinds (e.g., a set of bar charts showing the acreage of different crop types growing in an area in a particular season).

Factors of Pollution

The places from which pollutants emanate are called sources. Sources of pollutant gases may be natural, i.e., volcanoes, forest fires, or anthropogenic, i.e., automobile exhaust, etc.

The places to which pollutants disappear from the air are called sinks. Sinks include the soil, vegetation, structures and water bodies particularly the oceans. The mechanisms where by pollutants are removed from the atmosphere are called scavenging mechanism, and the measure, used for the aging of a pollutant is its half life–the time it takes for half of the quantity of pollutant emanating from a source to disappear in to its various sinks. For most of the pollutants it is in days.

A receptor is something which is adversely affected by polluted air. A receptor may be a person or animal that breathes the air and whose health may be adversely affected, or a plant that dies or some material like paper, stone or cloth that is affected. Transport is the mechanism that moves the pollution from a source to a receptor. Anthropogenic sources of air pollution.

No	Class	Aerosols	Gases and vapours
1.	Combustion processes (Cooking, heating, automobiles, thermal power plants, refuse burning)	Smoke, dust	SO_2, NO_2, CO, Organic vapours
2.	Chemical processes (paper mills, cement, fertilizers, etc.)	Dust, mist	Process dependent (SO_2, CO, NH_3, NO_2, Organic vapours, odours)
3.	Petroleum industry	Dust, mist	SO_2, H_2S, NH_3, CO, hydrocarbons, mercaptans
4.	Metallurgical processes	Dust, fumes	SO_2, CO, fluorides, organic vapours
5.	Mineral processing	Dust, fumes	SO_2, CO, fluorides
6.	Food industry	Dust	Odours
7.	Agricultural activities	Dust, mist Smoke	Pesticides, SO_2
8.	Nuclear energy systems	Dust	Fluorides, I-131 Sr-90, Cs-137, C-14, etc.

Pollutants of Primary and Secondary Nature: Air pollutants can be broadly classified into two general groups–primary air pollutants and secondary air pollutants.

Primary Air Pollutants: Primary air pollutants are those emitted directly from identifiable sources. Ex.

1. Fine particles (< 100 m in diameter) like pollen, bacteria
2. Coarse particles (> 100 P in diameter)
3. Sulphur compounds like SO_2
4. Oxides of nitrogen
5. Carbon monoxide
6. Halogen compounds
7. Organic compounds
8. Radioactive compounds

Secondary Air Pollutants: Secondary air pollutants are those which are produced in the air by the interaction among two or more primary pollutants, or by reaction with normal atmospheric constituents, with or without photo activation Ex.

1. Ozone
2. Formaldehyde
3. PAN (Peroxy acetyl nitrate)
4. Photochemical smog
5. Acid mist like H_2SO_4

Air Pollutants' Sampling: The sampling type, time, and size (volume of air sampled) varies with the nature and concentration of pollutant and sensitivity of the method.

Particulates: Filtration is the most common technique for sampling particulate matter. The filter material may be:

(a) Cellulose filter paper-for determination of metals
(b) Glass fibre-for analysis of organic compounds
(c) Silica felts for trace inorganic species and organic compounds. The optimum pore size of these filters is 0.45 m.

High volume air samples are used to pump large volumes of air up to 2000 m^3, at a rate of about 1.7 m^3/min. Usually glass fibre filter paper is used in it having a collection efficiency of more than 99% for particulates with 0.8 m diameter.

For workplace environmental monitoring light weight battery-operated samples are used which pump air at 2L min^{-1} through a small glass fibre filter paper for upto 8 hours. For particulate sampling other techniques used are sedimentation, cascade impactor and Electrostatic precipitation.

Gases and Vapours: The most common method of collection of gaseous pollutants is absorption in a solvent, by bubbling the gas through a liquid. The type of liquid medium depends on the nature of pollutant desired. Ex Alkaline solution for acidic gases, acidic solution for alkaline gases, oil for hydrocarbon, etc. The absorbers used are of various types e.g. impingers, fritted glass, packed column, etc.

All air pollutants from an air stream may be removed by cold trapping, i.e., freezing or liquefying them (gases and vapours) in collectors maintained at decreasing temperatures up to - 196°C.

Adsorption on charcoal, Al_2O_3, silica gel, etc. is used for collection of specific air pollutants.

Monitoring

Sulphur Dioxide: The modified West-Gaeke spectrophotometric method is the standard method for monitoring of 0.005-5 PPM SO_2 in ambient air. In this method SO_2 is collected in a scrubbing solution containing $HgCl_4^{2-}$ (Hg Cl_2+ KCI). The solution is allowed to react with HC HO and then with acidic pora-rosaniline hydro chloride to develop colour.

$$HgCl_4^{2-} + SO_2 + H_2O \rightarrow HgCl_2\ \ 50_3^{2-} + 2H^+ + 2Cl^-$$

$$SO_2 + H_2O + HCHO \rightarrow HOCH_2\text{-}SO_3H$$

p-rosaniline chloride + $3HOCH_2 - SO_3H \rightarrow$ Redviolet dye + $3H_2O + 3H^+$ (Colourless in acidic medium)

The absorbance of the product redviolet dye is measured at 548 nm. The major interference from NO_2 (> 2ppm) when present is eliminated by addition of Sulphonic acid which acts as a reducing agent converting NO_2 to N_2-

Air Pollutant's Impact on Humans: An average man breathes 22,000 times a day and takes in 16 kg of air each day. It far exceeds the consumption of food and water.

The prime factors deciding the effect of air pollutants on human health are:

(a) Nature of the pollutants

(b) Concentration of the pollutants

(c) Duration of exposure

(d) State of health of receptor

(e) Age group of the receptor

Air pollution mostly affects the respiratory, circulatory and olfactory system. The respiratory system is the principal route of entry for air pollutants.

Effects of Air Pollutants and their Health

Pollutant	*Health effects*
CO	Reduction in the ability of the circulatory system to transport O_2, impairment of performance on tasks requiring vigilance, aggravation of cardiovascular disease, high mortality rate
NO_2	Increased susceptibility to respiratory pathogens. Eye and nasal irritation at 15 ppm cone and pulmonary discomfort after brief exposure to 25 ppm cone.
O_1	Decrease in lung function, coughing, chest discomfort, increased asthma attacks
SO_2	Increased prevalence of chronic respiratory disease, bron-chospasm
Peroxyacylnitrates	Eye irritation
Formaldehyde	Eye and respiratory irritation
HF	Extremely irritant
Fluorine	Cumulative poison under conditions of prolonged exposure and low concen-trations.
Lead	Neurocognitive and neuro-motor impairment, Heme synthesis and hematologic alterations. Intestine, liver and kidney damage, abnormalities in fertility and pregnancy
Particulate matter	Lung problems

Impacts of Air Pollutants on Animals: In animals poisoning by air pollutants is a two-step process: accumulation of air-borne contaminants on or in vegetation or forage and subsequent poisoning of animals when they eat the contaminated vegetation. The effects can be divided into two parts.

Heavy Metal Effects: Arsenic poisoning of livestock has been found near smelters. Arsenic is also found as an impurity in many ores and coal. When arsenic trioxide concentrations. In feed is approx. 10 mg/kg body weight then toxic effects occur. In acute arsenic poisoning the symptoms are severe salivation, thirst, vomiting, uneasiness, feeble and irregular pulse and respiration. There is diarrhoea, and the feces have a garlic odour.

Atmospheric lead contamination is by smelters, coke ovens and coal combustion processes. Spray of lead arsenate can also contaminate plants. Cattle feeding on herbage containing 25-50 ppm lead develop excitable jerking of muscles, frothing at the mouth and paralysis of the larynx muscles.

Cattle feeding on vegetation containing 230 ppm of molybdenum are affected by molybdenum poisoning. Its symptoms are emaciation, diarrhoea, anaemia, stiffness and fading of hair colour.

Gaseous and Particulate Matter Effects: Phosphate fertilizer or phosphate derivative industries release fluoride as pollutants. Phosphate rock can contain upto 4% fluoride. Chronic fluoride toxicity (fluorosis) is the type most frequently observed in cattle. The primary effects of fluorides in cattle are seen in the teeth and bones. The common symptoms are mottling, staining and weasing of teeth, bony overgrowth on the skeleton, stiffness and lethargy and general ill health from malnutrition and starvation. It has been found that increased consumption of aluminium and calcium salts can reduce the toxicity of fluoride in animals.

Impacts of Air Pollutants on Materials: Air pollution damage to materials is a very important economic aspect of pollution. Air pollutants cause damage to materials by following mechanisms:

(a) Direct chemical attack e.g., tarnishing of silver by H_2S

(b) Indirect chemical attack e.g., conversion of SO_2, to H_2SO_4 before attacking leather

(c) Corrosion e.g., facilitation of metal oxidation by presence of air pollutants

(d) Abrasion or deposition by solid particles.

(a) *Effects on Metals:* The main effects of air pollutants on metals are corrosion of the surface, with eventual loss of material from the surface, and alteration in the electrical properties of the metals. The rate of corrosion of metals is influenced by moisture, type of pollutant and temperature. High moisture content and high temperature of the atmosphere accelerate the corrosion. Air pollutants can affect steel, aluminium, copper, zinc, brass and nickel. SO_2 is the main culprit for metallic corrosion.

(b) *Effects on Building Materials:* Building materials are corroded and disfigured by air pollutants. SO_2 and moisture react with limestone ($CaCO_3$) to form Calcium sulfate ($CaSO_4$) and gypsum ($CaSO_4$-$2H_2O$). These two sulfates are fairly soluble in water, causing deterioration of stone blocks and in the mortar used to hold the blocks together. CO_2 in the presence of moisture forms carbonic acid. This acid converts the limestone into bicarbonate which is water soluble and can be leached by rain. Marble structures like Taj Mahal are threatened by such type of deterioration.

(c) *Effects on Fabrics and Dyes:* Soiling and loss of tensile strength are the major effects of air pollution on fabrics. Sulfur oxides cause the greatest loss of tensile strength. Cotton, linen, rayon and nylon are damaged by SO_2 exposure. Dyes used to colour fabrics fade due to exposure to air pollutants. NO_2 and SO_2 are responsible for this fading of dyes. High moisture content increases fading.

(d) *Effects on Leather, Paper, Paint and Glass:* Sulfur dioxide affects the composition of leather and paper, leading to their deterioration. SO_2 is converted into sulfuric acid in leather and paper, which causes cracking in leather and brittling in paper. Leather can be protected by keeping it in SO_2 free air. Metal-free variety of paper is more resistant towards SO_2 deterioration.

Exposure to 1-2 ppm SO_2 increases the drying time of newly-painted surfaces. Hydrogen sulfide reacts with lead base pigments and blackens white and light-tinted paints. Sulfur dioxide and ozone also affect paints adversely. Hydrogen fluoride gas can affect glass which is normally a very stable material. Window glasses become opaque due to HE. It has been reported that SO_2 can affect old medieval glass by accelerating their deterioration.

(e) *Effects on Rubber:* In USA cracking of rubber is common due to presence of ozone in air produced by photochemical reactions. Ozone attacks the C-C double bound in rubber, thus cracks appear. In Los Angeles anti-ozone compound is added in tyres to prevent cracking.

Impact of Air Pollution on Plants: Air pollution has long been known to have an adverse effect on plants. There are examples where air pollution has completely destroyed the vegetation in the affected area.

The leaf tissue is in layers with a skin or epidermis layers on top and bottom and the photosynthetic cells in between. Stomata are the entrances in leaf through which gases enter or leave the stomata.

Damage to leaves by air pollutants is manifested in following forms:

1. Chlorosis	Loss or reduction in the green pigment, chlorophyll which results in pale green or yellow pattern.
2. Necrosis	Killing or collapse of tissue
3. Abscission	Shedding of leaves from the plant
4. Epinasty	Downward curvature of the leaf due to higher rate of growth on the upper surface.

2

Plant Protection

- Plant Pathology or Phytopathology (phyton-plant; pathos-ailments (disease); logos-knowledge) is that branch of agricultural science which deals with the cause etiology, resulting losses and management of plant disease. The science of plant pathology has four major objectives:
    1. To study the living, non-living and environmental causes of disease or disorder in plants.
    2. To study the mechanism (s) of disease development by pathogens.
    3. To study the interaction between the plant and the pathogen in relation to the overall environment.
    4. Develop systems of management of the diseases and reduce the loss of caused by plant diseases.
- Its aim is not only to destroy the enemies of plants but also to increase productivity of the crop and profits of the farmers.

Importance of Plant Ailments

Plant disease are important because of the loss they cause to the grower. The loss can occur in the field or in the store and at any time between sowing and consumption of the harvest standing crop in the field is attacked by a disease and plants start drying or their capacity to yield satisfactorily is reduced and sometimes creating conditions of food shortage and famine.

The late blight of potato, a disease caused by the fungus *Phytophthora infestance) is* a famous example of what a plant disease can do to change the course of history in 1845. This disease destroyed the potato crop of Ireland where potato constituted the staple diet of the majority in rural areas with Ireland, England and certain parts of continental Europe. Where potato was the staple food of population. There was famine in these countries especially Ireland and about 2 million people died of hunger and many more became diseased due to physical weakness. Many migrated to other land like North-America.

Wheat rust has been another disease that has appeared in epidemic form from time to time in many countries. This disease has forced the farmers in many part of the world to change their cropping pattern. Wheat has been replaced by corn (Maize) or rye because it was regularly destroyed by rust. In the year of the second world war (1943) Bengal had to face a serious famine due to attack of *Helminthosporium* leaf spot *(Helminthosporium oryzae)*. Likewise, in the middle of nineteenth century coffee rust. *(Hemileia zeastatrix)* attack the coffee plantation in Ceylon (Sri Lanka) which was the biggest producer and exporter of coffee. During 1867-71 the coffee production was reduced to almost half and by *1893* the export was reduced to *93%*. The planters were forced to cut down coffee plants and shift to tea plantation.

These are only few examples, however, ordinarily, plant disease affect every crop, every year and in all parts of the world. This result in huge loss of potential production and money. Accurate data regarding the losses are not available from all countries because of difficulties in assessment but on the basis of number of plants affected and extraordinary decrease in yield.

In India the losses due to rusts of wheat can range from 8-10%, loose smut of wheat is estimated to cause 2-3% losses. Few years back ear cockle of wheat, a nematode disease, was reported to cause a loss of 50% in U.P. Different smut of Jowar, red rot, potato mosaic, sandal spike disease, bunchy top of banana, rice blast, blight, etc. have caused huge losses in India.

Reasons of Plant Diseases: A pathogen is always, associated with a disease. The word "Pathogen" can be broadly defined as any agent or factor that incites "Pathos" or disease in an organism. Thus in strict sense the pathogens do not necessarily belong to living or animal groups. They may be non-living or in between the living and non-living. The pathogens are thus grouped under the following categories:

(a) *Abiotic Factors:* These include mainly the deficiencies or excess of nutrients, light, moisture, aeration, abnormalities in soil conditions, atmospheric impurities, etc. Examples of diseases caused by abiotic factors are: mango tip rot or fruit necrosis, "Khaira" disease of rice, hollow and black heart of potato.

(b) *Mesobiotic Causes:* These are disease incitants which are neither living nor non-living. They are considered to be on threshold of life.

(1) *Viroids:* Naked infectious strands of nucleic acid-spindal tuber of potato, citrus exocortis.

(2) *Viruses:* These are infectious agents made up of single type of nucleic acid (RNAor DNA) enclosed in a protein coat;

Instances: leaf curl of tomato and chilli, etc.

(c) *Biotic Causes:* This category includes diseases caused by animate or living or cellular organism.

(1) Prokaryotes

(a) *Mollicutes:* These are wall less prokaryotes that include mycoplasma-like organism (MLOs) and spiroplasmas.

Instances: Grassy shoot of sugarcane, little leaf of brinjal, sandal spike, papaya bunchy top.

(b) *Rickettsia like Bacteria (RLBs):* These are very small, sometimes sub-microscopic, walled bacteria causing such diseases as citrus greening.

(c) *True Bacteria:* Example of disease will of potato, soft rot of potato, citrus canker, etc.

(2) Eukaryotes

(a) *Fungi:* Potato wart, cabbage club root, potato late blight, downy mildews and powdery mildews, rusts and smuts, red rot of sugarcane, etc.

Nearly 70% of disease in any plant are incited by fungi.

(b) *Protozoa:* Hartrot of coconut palm, phloem necrosis of coffee.

(c) *Algae:* Red rust of mango, papaya.

(d) *Nematodes:* Root-knot of vegetables crops, molya disease of wheat and barley, ear cockle of wheat, citrus decline, etc.

Classification of Plant Ailments

These diseases may be infectious or non-infectious.

(1) *Non-infectious or Non-parasitic Diseases:* These are diseases (disorder) with which no animate, virus or viroid pathogen is associated. Therefore, they remain non-infectious and can not be transmitted from a

diseased plant to another plant. Since no parasite is associated with these disease they are also known as non-parasitic diseases.

(2) *Infectious Diseases:* These diseases are always infectious, sometimes contagious and are transmitted from diseased to healthy plants. These diseases are caused by parasitic pathogens or viruses.

Infectious diseases are often classified according to their occurrence in following groups:

(a) *Endemic Disease:* The word "endemic" means prevalent in and confined to, a particular country, district or location and is applied to diseases of plants and animals including man. These diseases are natural to one country or part of the earth, when a disease is more or less constantly present from year to year in a moderate to severe form, in a particular country or part of earth it is called endemic to that area.

(b) *Epidemic or Epiphytotic Disease:* The term "epidemic" is derived from a Greek word meaning "among the people" and in the true sense applies those diseases of plants humans/animals which appear very virulently among a large section of the population.

To carry the same sense in the case of plant disease the term "epiphytotic" was coined. An epiphytotic disease is one which occurs widely but periodically. It may be present constantly in, the locality but assumes serious proportions only on occasions.

(c) *Sporadic Disease:* Sporadic diseases are those diseases which occur at very irregular intervals and locations and in relatively few instances.

- When a disease is prevalent throughout the country, continent or the world it is known as endemic disease.

- For practical purpose the diseases are also classified according to their symptoms.
- According to symptoms the disease may be rusts, wilts, smuts, blights, cankers, mildews, root rot, fruit rots, leaf spots, etc.
- According to host plants the diseases may be grouped as cereals disease, forage crop disease, flax diseases, root crop diseases, plantation crop diseases, etc.
- When a pathogen survives and spreads through soil, it is known as soil-borne disease.
- However when a disease spreads through seed. It is known as seed-borne disease.
- When the disease dispersal through the agency of air the pathogen is known as air-borne disease.

As is discussed in the introductory para, the primary aim of the plant pathologist is to control the disease and increase the productivity of the crop. In order to devise suitable control measures one has to throughly study the cause of plant disease its etiology and factors favourable for its dispersal, infection and Epidemiology. Once the life cycle of the pathogen is completely understood, the pathologist would like to hit at the weakest link of its life cycle through immunisation or prophylatic means to achieve its control.

Plant Disease Control: Its Principles: Control practices are desirable only when the cost in terms of money and efforts is materially less than the loss expected from the disease so that the grower is ensured a margin of profit from plant protection measures since economically and ecologically it is not possible to completely eliminate a pathogen from the ecosystem, therefore, the word "control" is substituted with the word management. The disease management or control involves the following steps.

(a) Immunisation

(Possession of qualities by the host that do not permit establishment of infection).

1. *Resistance:* Development of resistant varieties.
2. *Chemotherapy:* Development of biochemical resistance through the application of antibiotics and systemic fungicides and also through modification of host nutritions.

(b) Prophylaxis (Preventive Measures)

(1) *Protection:*

(a) *Mechanical:* By removing the diseased material or seed from the healthy lot before sowing.

(b) *Physical:* By heat treatment, etc.

(c) *Cultural:* By modifying cultural operations like time of sowing depth, crop rotation, modification of environment, sanitation, roguing of diseased plants, production of disease free seed or propagated material.

(2) *Eradication:*

(a) *Crop Rotation:* Against those pathogens which persist for longer in the soil.

(b) *Sanitation:* To remove the diseased plants or residue from the field.

(c) *Alternate Host Eradication:* To eliminate the primary inoculum produced on them.

(d) *Plant Therapy:* Through fungicides heat and also surgery (in case of tree).

(e) *Chemicals:* Through fungicides and antibiotics.

(f) *Biological Control:* Through the activity of other micro organisms.

(3) *Legislation:* Aims at preventing entry of pathogen from infested area to non-infested areas.

Such regulations are called quarantine. The governments may have international or domestic quarantine against a particular disease.

Integration of Plant Disease Management Practices (IPDMP): The four basic requirement for management of plant diseases are: clean and healthy seeds, clean field or pathogen free soil, prevention of entry of and infection by a pathogen in a standing crop and precautions during harvesting and storage of the produce. The crop rotation is another precaution during growth of the crop periodical roguing of disease plant.

Plant Pathology Facts

1. Father of Pathology	Anton de Berry
2. Father of Mycology	P. A. Micheli
3. Father of Nematology	N. A. Cobb
4. Father of Indian Pathology	E. J. Butler
5. Bacterial technique was given by	Dr. Robert Koch
6. Mycoplasma disease was first reported by	Do. *et.* al. and Ishiil *et. al.*
7. First plant parasitic bacteria was reported by	T. J. Burill
8. First plant parasitic bacteria was	Fire blight of apple
9. First plant parasitic nematode was reported by	Needham
10. First plant parasitic nematode was	Anguina *tritici*
11. "Stanley" did crystallisation of	Viruses
12. Late blight of potato caused by	*Phytophthora infestance*

13. Brown spot of rice caused by	*Helminthosporium oryzae* caused Bengal femine (1845)
14. Coffee rust occurred in	Sri Lanka 1867
15. Maximum number of plant parasitic	Tylenchida order nematode belongs
16. Loose smut of wheat is	Internally seed borne *(JRF 2000)*
17. Indian phytopathological society was established by	E. J. Butler
18. Antibiotic penicillin was discovered by	Alexander Flemming
19. Example of obligate parasite is	*Albugo candida*
20. Green ear disease of bajra is a	Soil and air borne diseases
21. Life cycle of wheat rust is given by	K. C. Mehta
22. Systemic fungicide is discovered by	Ven Schleming and Kulka
23. Yellow vein mosaic virus was reported by	Uppal *et. al.*
24. Tomato mosaic virus was reported by	Adolph Mayer
25. Taphrina causes	Peach leaf curl
26. Endoparasitic nematodes is	Meloidogyne Lichen (Algae + Fungi).
27. Example of symbiosis is	Mycorrhiza (Fungi + Tree roots)
28. Citrus canker	*Xanthomonas campestris var. citri*
29. Molya disease of wheat due to	Nematode

30. Black tip of mango is caused by	Boron deficiency
31. Botulism is caused by species of	*Clostridium*
32. Agar-agar is produced from	*Gelidium*

Different Diseases of Plant

Bacterial Diseases

	Disease	*Causal Organism*
1.	Ratoon stunning disease of sugarcane (R.S.D.S)	*Cleibactor xyli*
2.	Stalk rot of maize	*Erwinia chrysanthemi pv. zeae*
3.	Tuber soft rot	*E. cartovora*
4.	Common scab of potato	*Streptomyces scabis*
5.	Citrus greening disease	*Diaphorina citri*
6.	Leaf spot of mango	*Xanthomonas campestris* pv. *Mangifera indica*
7.	Bacterial spike blight	*Clavibacter tritici*
8.	Bacterial brown rot of potato	*Pseudomonas solanacearum*
9.	Bacterial leaf streak of rice	*Xanthomonas campestris pv. oryzicola*
10.	Bacterial leaf blight	*Xanthomonas campestris pv. oryzae*
11.	Citrus canker	*Xanthomonas campestris pv. citri*
12.	Angular leaf spot	*Xanthomonas campestris pv. malvacearum*

Viral Diseases

1.	Leaf curl of papaya	*Nicotiana virus-10*

Contd...

	Disease	*Causal Organism*
2.	Papaya Mosaic	Potex virus group
3.	Fiji disease of sugarcane	Fiji virus
4.	Mosaic of sugarcane	Sugarcane Mosaic Virus (S.M.V.)
5.	Bunchy top of banana	Banana virus-1
6.	Leaf roll of potato	Potato leaf roll virus (Luteo group of virus) *(IAS(Pre) 2002)*
7.	Mild mosaic of potato	Potato mild mosaic virus
8.	Yellow vein mosaic of okra	Yellow vein mosaic virus
9.	Rugose mosaic of potato	Potato virus x and potato virus *y*
10.	Tomato spotted wilt	Tomato spotted wilt virus (Tospo group of virus) *(IAS (Pre) 2002)*
11.	Rice Tungro	Rice tungro virus (Bacilliform virus) (Badna group) *(IAS (Pre) 2002)*

Nematode Diseases

1.	Ear cockle of wheat	*Anguina tritici* *(HAS(Pre) 2003)*
2.	Molya disease of barley	*Heterodera avenae* *(PG. Bikaner 2000)*
3.	Root knot of vegetables crops	*Meloidogyne javanica* *M. incognita* *M. arenaria*

Contd...

	Disease	Causal Organism
Fungal Diseases		
1.	Club root of crucifers	*Plasmodiophora brassicae*
2.	Black wart of potato	*Synchytrium endobioticum*
3.	Damping off of seedlings	Most common generas *Phythium* *Phytophthora* *Fusarium* *Rhizoctonia* *Botrytis* *Ozonium*
4.	Brown spot disease of maize	*Physoderma zeae-maydis*
5.	Fruit rot of cucurbits	*Phythium aphanidermatum*
6.	Foot or stem rot of papaya	*P. aphanidermatum*
7.	White rust of crucifers	*Albugo candida*
8.	Green ear disease of bajra	*Sclerospora graminicola*
9.	Downy mildew of Jowar	*Sclerospora graminicola pr. andropopognis*
10.	Downy mildew of maize	*Pernosplerospora sacchari*
11.	Downy mildew of pea	*Pernospora pisi*
12.	Downy mildew of cucurbits	*Psedopernospora cubensis*
13.	Downy mildew of crucifers	*Pernospora parasitica* (PG 2000)
14.	Downy mildew *of* grape vines	*Plasmospora viticola*
15.	Soft rot of ginger	*P. myriotylum*
16.	Root rot of turmeric	*Pythium aphanidermatum*

Contd...

	Disease	***Causal Organism***
17.	Late blight of potato	*Phytophthora infestance*
18.	Blight of Colocasia	*Phytophthora colocasiae*
19.	Seedling blight of castor	*Phytophthora parasitica*
20.	Bud rot of palm	*Phytophthora palmivora*
21.	Gumosis of citrus	*Phytophthora palmivora and P. parasitica*
22.	Blight of sesamum	*P. parasitica var. sesami*
23.	Stem galls of coriander	*Promyces macrosporus*
24.	Peach leaf curl	*Taphrina deformis*
25.	Leaf spot of turmeric	*Taphrina maculans*
26.	Soft rot of apples	*Penicillium expansum*
27.	Powdery mildew of pea	*Erysiphi polygoni*
28.	Powdery mildew of cereals	*Erysiphe graminis*
29.	Powdery mildew of cucurbits	*E. cichoiacearum*
30.	Powdery mildew of grape vine	*Uncincola necator*
31.	Powdery mildew of apple	*Podasphara leucotricha*
32.	Ring spot of sugarcane	*Leptospharia sacchari*
33.	Stem rot of paddy	*Sclerotium oryzae*
34.	Ergot of cereals	*Calviceps purpuria*
35.	Ergot of bajra	*Calviceps fusiformis*
36.	False smut of rice	*Claviceps oryzae sativae*
37.	Apple scap	*Venturia inacqualis*
38.	Stem canker of fruit	*Necteria cinnabarinatrees*

Contd...

	Disease	Causal Organism
39.	Stem rot of papaya	*Sclerotinia sclerotiorum*
40.	Loose smut of wheat	*Ustilago tritici U. nuda tritici* (PG 2000)
41.	Brown rust of wheat	*Puccinia recondita*
42.	Yellow rust of wheat	*P. striiformis*
43.	Black or stem rust of wheat	*P. graminis tritici*
44.	Flag smut of wheat	*U. agropyri*
45.	Leaf smut of rice	*Entyloma oryzae*
46.	Bunt of rice	*Neovossia horrida*
47.	Karnal bunt of wheat	*Tilletia tritici*
48.	Stinking smut of wheat	*Tilletia caries*
49.	Smut of sorghum	*Tolyposporium ehrenbergii*
50.	Smut of bajra	*T. penicillariae*
51.	Head smut of sorghum	*Sphacelotheca reiliana*
52.	Grain smut of sorghum	*Sphacelotheca sorghi*
53.	Smut of maize	*Ustilago maydis*
54.	Covered smut of barley	*Ustilago hordi*
55.	Loose smut of bajra	*Ustilago nuda* (JRF1999)
56.	Smut of sugarcane	*Ustilago scitaminea*
57.	Covered smut of oats	*Ustilago kalleri*
58.	Loose smut of oats	*U. avenae*
59.	Bean rust	*Uromyces phoseoli typica*
60.	Cowpea rust	*U. phaseoli vignae*
61.	Leaf rust of coffee	*Hemelia vastatrix* and *Hemellia coffcicola*

Contd...

	Disease	Causal Organism
62.	Rust of linseed	*Melampsora lini*
63.	Early blight of potato	*Alternaria solani*
64.	Alternaria leaf spot of crucifers	*Alternoria brassicae*
65.	Leaf blight of wheat	*Alternaria triticina*
66.	Leaf spot of groundnut Tikka disease	*Cercasporidium personatum or C. arachidicola* (PC.2000)
67.	Leaf spot of rice	*Helminthosporium oryzae*
68.	Blast of rice	*Pyricularia oryzae*
69.	Red rot of sugarcane	*Colletotrichum falcatum*
70.	Die back of chillies	*Colletotrichum capsici*
71.	Mango anthracnose	*Colletotrichum gloeosporioides*
72.	Anthracnose of bean	*Colletotrichum lindemuthianum*
73.	Wilt of pigeon pea	*Fusarium udum* (PC 2000)
74.	Wilt of cotton	*Fusarium oxysporium sp. vasinfectum*
75.	Wilt of linseed	*F. oxysporium (f. lini)*
76.	Panama wilt of banana	*F. oxysporium (f. cubense)*
77.	Wilt disease of sugarcane	*Cephelosporium sacchari*
78.	Stalk rot	*Fusarium monilform*
79.	Blight of gram	*Ascochyta rabiei*

Contd...

	Disease	Causal Organism
80.	Charcoal rot of soyabean	*Macrophomina phaseolina*
81.	Sheath blight of rice	*Rhizoctonia solani*
82.	Black scurf of potato	*Rhizoctonia solani*
	Mycoplasmal Diseases	
1.	Little leaf of brinjal	Mycoplasma
2.	Purple top roll of potato	MLOs
3.	Marginal flavescence	MLOs
4.	Witches broom	MLOs *(JRF2002)*
5.	Grassy shoot of sugarcane	MLOs
6.	Sesamum phyllody	MLOs
	Physiological, Non-infectious or Non-parasitic Disease	
1.	Black heart of potato	Unfavourable O_2 condition during storage
2.	Black tip or mango necrosis	Polluted gases (SO_2)
3.	Tip burn of paddy	Reduced O_2 and N_2 under submerged condition
4.	Khaira disease of rice	Zn deficiency *(JRF1999;JRF2002)*
5.	Grey speck of oat	Mn deficiency
6.	Heart rot of beets	Lack of Mn
7.	Pahala blight of sugarcane	Mn deficiency *(IAS 2000)*
8.	White bud of maize	Lack of Zn
9.	Mottle leaf of citrus	Zn deficiency *(PCS 2002)*
10.	Whiptail *in* cauliflower	Mo deficiency *(JRF2001)*

Contd...

	Disease	Causal Organism
11.	Browning or hollow stem of cauliflower	Boron deficiency
12.	Top sickness of tobacco	deficiency of B
13.	Yellow spot of citrus	deficiency of Mo

Important Glossary

- *Abiotic Disease:* A disease due to non-parasitic cause.
- *Alternate Host:* One of the two kinds of plants on which a parasitic fungus must develop to complete the life cycle. A host of a pathogen growing during off season for the main host crop.
- *Antibiotic:* Chemical compound (Metabolite) of an organism the inhibits of other micro-organism..
- *Atrophy:* Under development or wasting away of a body part.
- *Autoecious Fungus:* A parasitic fungus that completes its entire life cycle on the same host.
- *Axenic:* That type of culture of an organism in which no other organisms are present.
- *Antheridium:* Male sex organ in fungi.
- *Bactericide:* A chemical that kills bacteria.
- *Bactericcin:* Bactericidal substance produced by certain strain of bacteria and active against one or more strains of the same or closely related species, their production is genetically controlled by extrachromosomal DNA.
- *Bacteriophage:* A virus that infects bacteria and usually destroys them.
- *Biological Control:* Total or partial destruction of population of a plant pathogen by use of other organism.

- *Biotechnology:* The use of genetically modified organisms and or modern techniques and processes with biological systems for disease control, crop production.
- *Blight:* Symptoms characterised by rapid killing of leaves, flowers and stems.
- *Blister:* A raised lesion on the leaf surface which opens to expose spores.
- *Blotch:* A disease characterised by large irregular spots or blots on leaves, shoots and stems.
- *Canker:* A necrotic, often sunken, lesion on stem, branch, twig or fruit.
- *Chlorosis:* Yellowing of normally green tissue due to destruction of chlorophyll or failure of chlorophyll formation.
- *Chronic Disease:* A disease which persists in the host for a long period of time.
- *Chronic Symptoms:* Symptoms that appear over a long time.
- *Contagious Disease*: A parasitic disease whose pathogen can spread from plant to plant by contact.
- *Circulative Viruses*: Viruses that are acquired by their vectors through their mouth parts, accumulate internally, then are passed through their tissues and introduced into new plants via. the mouth part of the vector.
- *Damage*: Any reduction in the quantity and/or quality of yield that results from injury.
- *Damping Off*: Seed decay in soil or seedling or blight usually caused by soil inhabiting fungi.
- *Diagnosis*: Identification of the nature and cause of the disease.

- *Die-back*: A symptoms of disease characterised by progressive death of shoots branches and roots from the tip backward.
- *Disease Assessment*: Quantitative measurement of disease or the process of measuring disease quantitatively.
- *Disease Intensity*: General term for amount of disease present in a population.
- *Disease Prevalence*: Generally incidence of fields with diseased plants in a defined geographic area.
- *Downy Mildew*: A plant disease in which the sporangio or conidiophores and spores of the fungus appear as downy growth on the host surface.
- *ELISA (Enzyme-Linked Immuno Sorbent Assay):* A serological test in which one antibody carries with it an enzyme that releases a coloured compound.
- *Epidemic:* A widespread and severe outbreak of a disease in a large population.
- *Epidemology:* The study of factors affecting outbreak and spread of an infectious disease.
- *Eradicant:* A chemical that destroys a pathogen at its source.
- *Eradication:* Control of plant diseases by eliminating the pathogen after it is established or by eliminating the plant that carries the pathogen.
- *Etiology:* Study of science of the cause of the plant disease and nature of the causal agent.
- *Exanthema:* Eruption or discharge of gum or other substances from diseased tissues.
- *Facultative Parasite:* An organism that is usually a saprophyte but which under some conditions may become a parasite.

- *Facultative Saprophyte:* An organism which spends most of its life as a parasite but may become a saprophyte for sometime under certain conditions.
- *Fumigant:* A volatile toxic substance or gas used to disinfest certain areas or plánts.
- *Fumigation:* The application of a fumigant for disinfestation.
- *Fungicide:* A compound that kills fungi.
- *Gall:* A swelling or overgrowth produced on a plant as a result of infection by certain pathogens.
- *Gummosis:* Production of gum by or in a plant tissue.
- *Homothallism:* The phenomenon in which a fungus produces compatable male and female gametes on the same mycelium.
- *Horizontal Resistance:* Partial resistance equally effective against all races of a pathogen.
- *Heterothallic Fungi:* Fungi producing compatable male and female gametes on physiologically distinct mycelia.
- *Hyaline:* Colourless, transparent.
- *Hypovirulence:* Reduced virulence of a pathogen stain as a result of the presence of transmissible double stranded RNA.
- *Infection:* The establishment of parasitic relationship of a parasite with the host.
- *Injury:* Damage to a plant by animal, physical or chemical agent visible or measurable symptoms and/or signs caused by a pathogen or pest.
- *Integrated Control:* An approach to plant disease management that uses all available methods of control of a disease or of all or most of the diseases or pests of a crop plant for best control results but with the least cost and the least damage to the environment.

- *In Vitro:* In culture, outside the host.
- *In Vivo:* In the host.
- *Intracellular:* Within the cells through the cells.
- *Intercellular:* Between the cells.
- *Leaf Spot:* A self limiting lesion on the leaf.
- *Local Lesion:* A localised spot produced on a leaf upon mechanical inoculation with a virus.
- *Latent Virus:* A virus that does not induce symptom development in the host.
- *Latent Infection:* The state in which a host is infected but does not show symptoms.
- *Mycoplasma Like Organism (MLO):* The organism that other wise looks and behave like a mycoplasma but has not yet been grown in artificial culture.
- *Mottle:* An irregular pattern of indistinct light and dark areas.
- *Mosaic:* Symptoms of certain virus diseases of plants characteried by intermingled patches of normal and light green or yellowish colour.
- *Molt:* To cast off the cuticle; Eedysis instar.
- *Mold:* Any profuse or woolly fungus growth on damp or decaying matter or no surfaces of plant disease.
- *Mildew:* A fungal disease of plants in which the mycelium and spores of the fungus are seen as a whitish growth on the host surface. When the growth appears like a downy growth it is called downy mildew.
- When it is like a white or dull powder it is known as powdery mildew.
- *Necrotic:* Dead and discoloured.
- *Nematicide:* A chemical compound or physical agent that kills nematodes.

- *Non-infectious Disease:* A disease that is caused by an environmental factor, not by organisms and viruses.
- *Obligate Parasite:* A parasite that in nature can grow and multiply only on a living organism.
- *Oomycetes:* A fungus that forms oospores.
- *Parasite:* An organism that lives on or in another living organism and draws its nutrition from that organism.
- *Pathogen:* An entity that can incite disease.
- *Polygenic:* A character controlled by many genes.
- *Proliferation:* Abnormal outgrowth.
- *Phytotoxic:* Tox to plants.
- *Pathogenicity:* The ability of a pathogen to cause disease.
- *Postule:* A small blister-like elevation of the epidermis, often opening to expose spores.
- *Quarantine:* Control of import and export of plants or plant parts to prevent spread of disease and pest.
- *Resistance:* Ability of a host plant to supress or retard the activity of a pathogen or either injurious factors.
- *Resistant:* Possessing qualities that hinder the development of a given pathogen, little or no infection.
- *Ring Spot:* A disease symptom characterised by a circular area of chlorosis with a green centre often caused by viruses.
- *Rosette:* Short, bunchy habit of plant growth.
- *Rot:* The softening discolouration and often disintegration of a succulant plant tissue as a result of fungal or bacterial infection.
- *Rust:* A disease giving a rusty appearance to a plant and caused by one of the uredinales.

- *Scab:* A roughened crust-like disease area of the surface of a plant organ. A disease in which such areas are formed.
- *Scorch:* "Burning" of leaf margins as a result of infection or unfavourable environmental conditions.
- *Secondary Infection:* Infection caused by inoculum produced as a result of primary infection.
- *Serology:* A method using specificity of the antigen-antibody reaction for the detection and identification of antigenic substances and the organisms that carry them.
- *Smut:* A disease caused by smut fungi and characterised by appearance of shooty powdery mass on the affected organs.
- *Sooty Mold:* A shooty coating on foliage and fruits formed by the dark hyphae of some fungi that live on the honeydew secreted by insects.
- *Sporadic:* Term applied to a disease that breaks out occasionally without being constantly destructive.
- *Spore:* A single to many cell reproductive body of fungi and some lower plants which may develop into a new plant.
- *Stem-pitting:* A symptom of some virus diseases of trees characterised by depressions on the stem under the bark.
- *Sterile Fungi:* Fungi in which no form of spore has been seen.
- *Susceptibility:* The inability of a plant to resist the effect of pathogen or other injurious factors.
- *Symptom:* The external and internal reactions of the host to invasion of a pathogen alterations in the plant due to infection; signs of infection.

- *Syndrome:* A group of signs and symptoms that occur together and characterise the disease.
- *Systemic:* Spreading throughout the plant body internally.
- *Tolerance:* The ability of a plant to sustain the effects of a disease without dying or suffering serious injury or crop loss.
- *Toxicity:* The capability of a compound to produce injury.
- *Tumour:* And uncontrolled overgrowth of tissue or tissues.
- *Vertical Resistance:* Complete resistance to some races of the pathogen but not to other.
- *Virion:* A complete virus particle.
- *Viroids:* Small, naked, low molecular weight ribonucleic acid (RNA) that can infect plants, replicate, themselves and cause disease.
- *Virus:* A submicroscopic obligate parasite consisting of only nucleic acid and coat protein.
- *Virulent:* Capable of causing a severe disease strongly pathogenic highly aggressive.
- *Wilt:* Loss of rigidity and dropping of plant parts generally caused by insufficient water in plant body or by infection resulting in blockage of water transport or toxicity.
- *Witches Broom:* Broom like overgrowth or massed proliferation caused by dense clustering of branches of a plant.
- *Yellows:* A plant disease characterised by yellowing and stunting of the host plant.
- *Zoospores:* Spores bearing flagella and capable of movement in water.

Table

Crop	*Disease or Insect*	*Fungicide or Insecticide Comman Name*	*Fungicide or Insecticide Trade Name*	*Remarks*
Barley, Oats, Rye, Wheat	Seed decay and seeding blights including the scab fungus (seed treatments do dot control the head blight phase of scab disease)	Captan	Captan 30-DD Captan 300 Captan 400 Captan 400 DD	Thiram formulations are not registered for oats. Vitavax formulations, Raxil-Thiram, add Terra-coat LT-2 are dot registered for rye. Use according to instructions on label. Do dot use treated seed for feed or food.
		Thiram	Thiram-50WP Gustafson 42-S	
		Corboxin	Vitavax	Not for use on rye
		carboxin + ca tad	Vitavax 20-20	
		carboxin + thiram	Vitavax 200	
		difenconazole + mefanoxam	Dividend XL RTA 5-10 fl oz/100	For wheat use only. Apply pounds of seed.
		difenconazole + mefanoxam	Dividend Apron XL 0.5-1.0 fl oz/100	For wheat use only. Apply pounds of seed.
		PCNB	Gustafson LT-2N	Not for use on rye
		tebuconazole + thiram	Raxil-Thiram	

Contd...

Barley	Loose smut, Covered smut	carboxin	Vitavax Vitavax 34 Vitavax 75 W	Use according to instructions on label. Must be applied as a slurry treatment for complete coverage.
		carboxin + ca tad carboxin + thiram tebuconzole + thiram	Vitavax 20-20 Vitavax 200 Raxil-Thiram	
	Covered smut	PCNB	Terra-Coat LT-2	Use according to instructions on the label. Should be applied as a slurry treatment for complete coverage.
	Barley stripe	imazalil	Gustafson Flo-Pro IMZ	Use according to label instructions
Oats	Loose smut, covered smut	carboxin carboxin carboxin + thiram	Vitavax Vitavax 34 Vitavax 200	Use according to instructions on label. Must be applied as a slurry treatment for complete coverage.
	Loose smut	PCNB + tebuconazole + thiram	Terra-Coat Lt-2 Raxil-Thiram	Use according to instructions on the label. Should be applied as a slurry treatment for complete coverage.

Contd...

Wheat	Loose smut	Carboxin	Vitavax Vitavax 34	Use according to instructions on label. Must be applied as slurry treatment for complete coverage.
		Carboxin + ca tan	Vitavax 20-20	
		carboxin + thiram	Vitavax 200	
		difenconazole + mefanoxam	Dividend XL RTA	Apply 5-10 fl oz/100 Ib of seed.
		defenconazole + mefanoxam	Dividend Apron XL	Apply 0.5-1.0 fl oz/ 100 Ib of seed.
		tebuconazole + thiram	Raxil-Thiram	
		triadimen	Baytan 30 Flo%vable	This material is only registered for application by certified seed conditioners and is not available for farmer application. Rate is 0.75-1.5 fl oz/ 100 Ib of seed. Caution:Treated seed should not be planted late in the planting season or planted more 1.5 inches deep. Use according to label instructions.

Contd...

Wheat	Common bunt or stinking smut	carboxin Vitavax 34	Vitavax instruction on label. Must	Use according to be applied as a slurry treatment for complete coverage.
		carboxin + captan	Vitavax 20-20	
		carboxin + thiram	Vitavax 200	
		PCNB	Gustafson LT-2N	
		tebuconazole +	Raxil-thiram thiram	
		difenconazole + mefanoxam	Dividend XL RTA	Apply 5-10 fl oz/100 Ib of seed.
		difenconazole + mefanoxam	Dividend Apron XL	Apply 0.5-1.0 fl oz/ 100 Ib of seed.
Wheat	Common root rot complex	imazalil	Gustafson Flo-Pro IMZ	Use according to label instructions.
		difenconazole + mefanoxam	Dividend XL RTA	Apply 5-10 fl oz/100 1b of seed.
		difenconazole + mefanoxam	Dividend Apron XL	Apply 0.5-1.0 fl oz/ 100 Ib of seed.

Contd...

Wheat	Powdery	tridimenol	Baytan 30 Flowable	This fungicide will prevent the mildew overwintering of powdery mildew and leaf rust in susceptible cultivars. Rate is 1.5 fl oz/100 Ib of seed. It is only registered for available for farmer application. Caution: Treated seed should not be planted late in planting season or planted more than 1.5 inches deep. Use according to label instructions
Wheat	Leaf rust	difenconazole + mefanoxam	Dividend XL RTA	Apply 5-10 fl oz/100 Ib of seed.
		difenconazole + mefanoxam	Dividend Apron XL	Apply 0.5-1.0 fl oz/ 100 Ib of seed.
Wheat barley	Aphids (Barley Yellow Dwarf) Hessian fly	imidazolidinimine	Gaucho 480 Flowable	Apply 1.0 fl oz/100 Ib of seed for fall aphid control. Apply 1.5-2.0 fl oz/100 Ib of seed for Hessian fly control.

Contd...

Leaf Rust	Apply fungicide if rust covers 1 to 3% of area of upper leaves
Powdery Mildew	Apply fungicide if powdery mildew covers 5 to 10% of area of upper leaves
Septoria	Apply fungicide if 25% of the indicator leaves (described below have one or more lesions.

Growth Stage	**Indicator Leaves**
Jointing until flag leaf emergence:	4th and 5th leaf below the flag
Flag leaf emergence until late boot:	3rd leaf below the flag
Late boot until start of flowering:	2nd leaf below the flag
Flowering until early milk stage:	1st leaf below the flag

Crop	Disease	Fungicide Common Name	Fungicide Trade Name	Formulated Rate/acre	Remarks
Wheat and Barley	Powdery mildew on wheat and barley	Propico-nazole	Tilt 3.6 EC	4.0 fl oz	Apply only when powdery mildew covers 5-10% of area of fully expanded upper leaves. Do not apply more than 4 fl oz/a per year or after Feekes' GS 8 (flag leaf fully emerged). Do not graze or feed livestock treated forage or cut he green crop for hay or silage. After harvest, straw may be used for bedding. Follow all label instructions.

Contd...

Wheat and Barley	Leaf rust on wheat and barley	Propic-o nazole	Tilt 3.6 EC	4.0 fl oz	Apply only when rust pustules cover 1-3% of area of fully expanded upper leaves. Do not apply more than 4 fl oz/a per year or after Feeke's GS 8 (flag leaf fully emerged). Do not graze or feed livestock treated forage or cut the green crop for hay or silage, after harvest straw may used for bedding. Follow all label instructions.
Wheat	Leaf Rust	azoxyst-robin	Quardris oz	6.2-10.8 fl	Integrate into an overall disease management strotegy that includes proper selection of varieties with disease tolerance, proper timing and placement of irrigation, removal of plant debris in which inoclum overwinters, plant residue management and crop rotation. Resistance management: Do not make more than two application of Quadris per acre per. Application directions Quadris should be applied prior to or in the early stages of disease development. Application may be made at any time, immediately after jointing (Feekes 6) up to late head emergence (Feekes 10.5) A crop oil concentrate adjuvant may be added at 1.0% V/V to optimise efficacy. Do not hervest treated wheat for Forage apply later than Feekes growth stage 10.5. Do not apply more than 0.77 quarts product/acre/season (0.4 lb ai/a). Do not apply within 14 days of harvest for hay. Do not apply within 45 days of harvest for rain and straw.

Contd...

Barley and rye	Barely scald, Net, blotch	Propic-onazole	Tilt 3.6 EC	4.0 fl oz	Do not graze or feed livestock treated forage or cut the green crop for hay or silage. After harvest, straw may be used for bedding. Follow all label instructions.
Wheat	Leaf and glume blotch and Tan spot	Propico-zole	Tilt 3.6 EC	4.0 fl oz	Tilt: Do not graze or feed livestock treated forage or cut the green crop for hay or silage. After harvest, straw may be used for bedding. Follow all label instruction.
Wheat	Leaf and glume blotch and tan sopt	azoxyst robin	Quardris fl oz	6.2-10.8	Quadris should be integrated into an ovreall disease management strategy that includes proper selection of varieties with disease tolerance, proper timing and placement of irrigation, removal of plant debris in which inoculum overwinters, plant residue management and crop rotation. ***Resistance Management:*** Do not make more than two application of Quadris should be applied prior to or in the early stages of disease development. Application may be made at any time, immediately after jointing (Feekes 6) up to late head emergence (Feekes 10.5) A crop oil concentrate adjuvant may be added at 1% V/V to optimise efficacy. Do not apply later than Feeks Growth Stage 10.5 Do not harvest treated wheat for forage. Do not apply more than 0.77 quarts product/acre/season (0.4 lb ai/a) Do not apply within 14 days of harvest for hay. Do not apply within 45 days of harvest for grain and straw.

3

Role of Insects

Lac insects and their products have been known to naturalists since very early times. The lac has been referred in ancient Sanskrit works, *viz.*, Atharva-Veda (Dave, 1950; Hora, 1952) and was called as 'Luxa'. It is mentioned in Mahabharata that 'Luxa Griha' was made up of lac which was prepared by Kaurava for Pandavas. Abul Fazal (1590) in his famous book 'Ain-i-Akbari' has mentioned in detail about the lac industry in India.

Mahdihassan (1950, 1952), has referred about the lac insect and its products in China. The first scientific reference regarding the lac and lac insect is the report of Kerr and Glover in 1782. Subsequently, much work has been done by various workers on the organisation, distribution, taxonomy, host plants, culture, production, enemies, chemistry and technology.

Three products from lac insects, *viz.*, the lac-dye, lac-wax and lac (resin) have been items of trade and commerce.

Distribution: India has its monopoly on the production of lac. Other countries like Africa, Australia, Brazil, Burma, Sri Lanka, China, Formosa, France, W. Germany, Japan, Malaya, Nepal, Spain, Thailand, Turkey, U.S.A. and some others also produce lac. But in Thailand, Malaya, Burma and Nepal the lac producing industries are increasing day-by-day. Thailand has become the main competitor of India in export of lac.

In India major lac producing places are Assam (Kashi Hills), Bengal (Calcutta, Jangipur, Murshidabad, Mathrapur, Malda), Jharkhand (Manbhum, Palamau, Ranchi, Santhal Pragana), Delhi, Gujarat, Hyderabad, Kashmir, Madhya Pradesh (Rewa, Umaria), Chattisgarh (Damoh, Champa, Bilaspur), Chennai, Coimbatore, Mysore, Orissa (Cuttak, Mayurbhanj), Punjab (Hoshiarpur, Shahpur), Rajasthan (Indergarh, Kota, Jaipur, Jhallawar, Karauli), and Uttar Pradesh (Ghazipur, Mirzapur, Agra), etc.

Lac Insect

Phylum .... Arthropoda
Class .... Insecta
Order .... Hemiptera
Sub-order .... Homoptera
Super-family .... Coccidae
Family .... Leciferidae
Genus .... *Tachardia*
Species .... *lacca*

Lac insect (*Tachardia lacca)* previously known as *Laccifer lacca* is a minute, resinous, crawling scale-insect which inserts its beak into plant tissues, sucks juices and grows, and secretes lac from the hind end of the body. Its own body ultimately gets covered with lac in the 'CELL'. Lac is actually secreted for its protection and not for the food of the insect. The commercial lac is produced in large quantities by female as a protective covering of its body which is injurious to the host plants.

Male: Male is red in colour and 1.2 to 1.5 mm in length. It secretes bright creamy lac. It has reduced eyes and ten segmented antennae. The mouth-parts are of piercing and sucking type. Thorax bears three pairs of legs and one pair of hyaline wings. The abdomen is eight segmented and terminates into a short, chitinous prominent general sheath containing penis. On either side of this genital sheath a white elongated caudal seta is found.

Female: Female is larger than males and measures about 4 to 5 mm in length. The pyriform body of the female is enclosed in a resinous cell. The head, thorax and abdomen are not clearly distinct. The mouth-parts are of piercing and sucking type. The antennae are clearly visible and degenerated. The posterior end of the body has a median and two lateral processes. The legs are in degenerated form.

The Background

Each mature female just after fertilization lays about 200 to 500 eggs in a cell in which she is enclosed. The oviposition takes place into the incubating chamber which is formed by the contraction of the body of the female in forward direction inside the lac cell. The eggs are laid in the months of October and November. After six weeks of laying, the eggs are hatched into first instar nymphs in the months of November and December. When nymphs emerge they are in quite large number. This mass emergence of the nymphs is known as 'SWARMING' (Fig.).

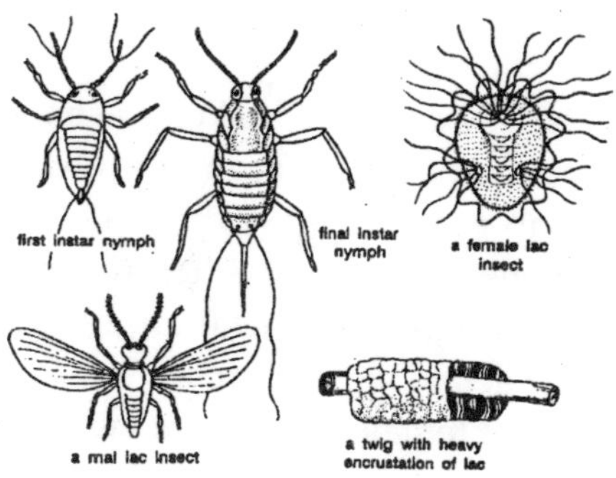

Life history of Tachardia lacca.

Nymph: At the time of emergence the nymphs are about 0.5 mm in length, red coloured and boat-shaped. The head bears paired antennae, ocelli and ventrally situated piercing and sucking type of mouth-parts. The mouth-parts are provided with proboscis. The three segmented well developed thorax contains two pairs of spiracles and only one pair of walking legs. The abdomen contains two pairs of legs and terminates into a pair of long caudal setae.

The active nymphs can crawl to a considerable distance so, just after emergence they start moving in search of food and reach their host plants, preferably on young and succulent shoots because the young nymphs are unable to settle and feed on hard twigs. These nymphs settle very close to each other on the twig of the host plant which further collapses completely and forms a continuous covering even on the lower surface of the twig.

The number of nymphs that settle per square inch area is about 150 to 200. Settled nymphs suck the sap from the twig of the host plant and start to secrete the resinous substance by special dermal glands which are located all over the body. As the resinous secretion comes in contact with air, it soon becomes hard and forms a coating over the body of nymph and is called as 'CELL'. Within this cell various life processes like growth of the nymph, morphological changes and lac secretion take place.

The male 'Cell' is elongated and cigar-shaped having two holes, *i.e.,* anterior and posterior. From the posterior hole which is covered by a flap or operculum, the male insect comes out by pushing open the operculum. After six to eight weeks of stationary life the nymphs are metamorphosed as a result of which some (30%) active winged males and maximum (70%) emerge in the form of females which are wingless. The females get fixed on the host plant in resinous mass. The males walk over the encrustations of females and fertilize them within their oval cells through anal opening. The males leave the parent cell after fertilizing the female. One male is capable to fertilize many females.

The female nymph once settled never moves but undergoes 3 moults inside her cell loosing its eyes and legs, and with rudimentary antennae only. The fertilization of female is followed by a rapid growth of the female body till it begins to lay eggs in October and November. From these eggs male and female emerge in February to March. The male fertilizes the females of this generation and the fertilized female lays eggs in months of June to July and dies secreting lac all the time. Thus, the life cycle reoccurs twice in one year on the same host plant.

Due to short life period males do not take major part in the secretion of lac but female secretes lac throughout her life and its life span is longer than males. Major quantity of lac is secreted from females. The life cycle period depends mainly on ecological factors of the region.

Supportive Plants

The lac insects have more than one host plant. The selection of suitable host plant for the cultivation of lac is of much importance. To establish the lac industry one should know well about the topographic and climatic conditions for the growth of host plants suitable for that particular region. Brun (1958) has mentioned that 113 varieties of host plants are found in the geographical Indian regions including Pakistan and Burma. Out of these 113 host plants only 14 are very common in India which are as follows:

1. Kusum — *Schleichera oleosa*
2. Babul — *Acacia nilotica*
3. Ber — *Zizyphus mauritianas*
4. Palas — *Butea monosperma*

5. Ghont — *Zizyphus xylopyra*
6. Khair — *Acacia catechu*
7. Peepal — *Ficus religiosa*
8. Gular — *F. glomerata*
9. Pakapi — *F. virens*
10. Putkal — *F. globella*
11. Mango — *Mangifera indica*
12. Sal — *Shorea robusta*
13. Shisham — *Dalbergia sisso*
14. Fig — *Ficus carica*

The quality of lac is directly related with the quality of host plant. So far, no artificial product has been able to replace the lac. Khair, Kusum and Babul give better quality of lac when sown directly in the field. But Palas, Ber and Ghont give good crop when they are first sown in nursery and then transplanted to the lac growing field. Palas and Ber produce a particular type of lac which is called as 'KUSUMI LAC'.

Cultivation of Lac: Lac cultivation is a complicated process, so the cultivators should know well about the inoculation, swarming period and harvesting of lac.

Inoculation: The first procedure in the lac cultivation is the inoculation of lac insect. Inoculation is the process by which young ones get associated properly with the host plant. Inoculation is of two types:

Natural Inoculation: The inoculation taking place in normal routine or in natural way is very simple and common process during which the swarmed nymphs infect the same host plant again and start to suck the juices from the twigs. The natural incubation of swarmed nymphs has some drawbacks which are as follows:

(a) *Incomplete Nutrition:* Lac insects with their piercing and sucking mouth-parts, pierce into succulent twigs and suck the cell sap of the same host plant for nutrition. If the cell sap of the same host plant is further sucked out by the swarmed nymphs of the second crop continuously, the growth of the host plant would be retarded. In this way lac insect may not be able to get enough nutrients from the same host plant. The lac insects due to lack of sufficient nutrients lose their proper development, thereby affecting the production of lac also.

(b) *Irregular Inoculation:* During the natural inoculation it is not sure that uniform sequence of inoculation takes place. If inoculation is not of continuous fashion, a regular crop of lac may not be obtained.

(c) *Unfavourable Climatic Conditions:* At the time of swarming a number of factors like high intensity of sunlight, heavy rainfall, flow of wind, etc. affect the proper inoculation of nymphs. These natural environmental factors may also affect the host plant at the same time and may cause a gap of inoculation resulting in irregularity of the lac crop.

(d) *Multiplication of Parasites and Predators:* Lac insects have certain enemies in the form of parasites and predators. If the crop is not harvested in time and lac is allowed to remain on the same twig, the multiplication of parasites and predators takes place which hampers the population growth of lac insects.

Thus, keeping in view the above drawbacks the natural procedure of inoculation is avoided and certain devices have been developed to ensure artificial method of inoculation.

Artificial Inoculation: The main idea behind the artificial method of inoculation is to check all possible drawbacks of natural inoculation.

In this method first of all host plant should be pruned in January or June. The twigs bearing insect nymphs which are about to swarm, or just before swarming are cut in sizes ranging between 20 to 30 cm in length. Then the cut pieces of these twigs are tied to fresh trees in such a way that each stick touches the tender branch of the tree at several places which form bridges for the migration of the nymphs. After swarming, these twigs should be removed and separated from the host plant. The following precautions should be taken in artificial inoculation:

(i) One must ensure that the twigs, which are going to be tied on fresh host plant, are having good number of nymphs or eggs. It is also possible that from many of the twigs nymphs have swarmed out, thus inoculation would prove unsuccessful.

(ii) The twigs provided with eggs or nymphs should be without any parasite and predator.

(iii) The eggs or nymphs present on the twigs should be healthy and about to swarm so that one has not to wait for longer period and thus save time.

(iv) For the uniformity of inoculation, 3 to 4 twigs should be utilised.

(v) Host plants should be changed from time to time for the proper nutrition of the nymphs.

These insects are very small and if they move to a long distance there are chances of mortality of the nymphs. Due to maximum contact of twigs, swarming nymphs have not to move for long distance and find suitable places to establish on the host plant.

Inoculation Period: In India two types of crops, *viz.*, Rangini and Kusumi are grown in a year. The Rangini crop is of two types called as Kartiki and Baisakhi crop which produce Kartiki and Baisakhi lac respectively. The Kusumi crop is also of two types, *viz.*, Agahani and Jethi which produce Agahani and Jethi lac respectively.

Thus, the inoculation periods of all the four types of crops are different. The inoculations of Kartiki, Baisakhi, Agahani and Jethi crops are recommended in months of June to July, October to November, July and January to February respectively. But if continuously four crops are taken, the plant would not get any rest which may cause less production of lac.

Swarming: It is very important phase in the life history of lac insect. So one should have accurate knowledge about the actual date of the swarming. At the time of swarming, the upper surface has yellow spot on the anal region. At this stage muscle contracts and insect gets detached from the place of attachment. Thus, it leaves a hollow cavity which later on gets covered with wax also. When these eggs are to be hatched out they become orange coloured. Thus, it is an indication that swarming has taken place. Thus, by trials and learning methods, *i.e.,* by practice one could know about the exact date of swarming by looking at the colour of the eggs.

Harvesting of Lac: The process of collection of ready lac from host tree is known as harvesting. In common practice the harvesting is of two types:

1. *Immature Harvesting:* The harvesting of the lac before swarming is called as immature type of harvesting and the lac thus obtained is known as 'ARI LAC'.
2. *Mature Harvesting:* The collection of crop after the swarming is called as mature harvesting and the lac obtained is known as 'MATURE LAC'.

The harvesting of lac before the swarming has some drawbacks because the lac insects may be damaged at the time of harvesting which would affect the population of lac insects and ultimately result in great economic loss to the cultivators. But in case of palas lac (Rangini lac) it is found that Ari lac gives better production. Therefore, Ari lac harvesting is recommended in case of palas only. In all other cases immature harvesting should be discouraged. It is also found that in cold areas mature crop yields better quality of lac.

Harvesting Period: The harvesting periods of different crops are quite different in accordance with the inoculation of crops. Kartiki crop is harvested in October to November whereas, Baisakhi crop in May and June. The other crops like Agahani and Jethi are harvested in January to February and June to July respectively.

Recent Plan for Lac Cultivation: With the increasing number of lac industries some advanced plans have been recommended for the better cultivation of lac crops. Two types of planning are used nowadays.

1. *Ceupe System:* All the trees of host plants of a definite area are not used under continuous cultivations process of lac crop because if all host plants of a farm would be under continuous attack of lac insects, 100% plants may not get any rest and thus the production of the lac would be affected due to deficiency of nutritive cell sap to the swarmed nymphs and adults. So, the plants of a farm are numbered into 5 groups of plants. This artificial division or marking of trees is called as ceupe system of crop cultivation. In this system when one group of host plants is under the process of cultivation of lac, other groups of host plants would be under rest.
2. *Alternation of Plant:* In this system the variety of host plant is changed after one crop. So, swarmed nymphs are inoculated on the tree of other variety of host plant. In this way every host plant can get enough rest resulting into better production of lac.

Processing of the Lac Industry: When the crop matures fully most of the lac is harvested and some part is left on the host plant. For the proper cultivation, the host plant should be pruned in January every year.

The twig bearing the lac alongwith eggs is called a BROOD LAC STICK and lac is known as BROOD or STICK LAC. The processing starts with the scraping of the stick lac from the twig. The scraped lac is subjected to removal of many impurities like dead parts of the lac insects, eggs and colouring matter, and finally crushed by hand-operated mortars.

Then the material is air dried and obtained in the form of granules which is known as SEED LAC. This seed lac is soaked in water, washed, dried in sun light, bleached and heated to melt on charcoal fire in cloth bag of 3 to 4 metre. At the time of heating the bag is twisted and the lac is squeezed out of the bag. The impurities of the lac are left out in the bag, and are called as KIRRI LAC.

The squeezed lac is now allowed to cool and solidify around the button-shaped forms which is now called BUTTON LAC or PURE LAC. This pure lac when stretched into thin sheet is called as SHEET LAC. This sheet lac when dissolved in water, produces white or orange coloured lac which is called as SHELL LAC. Shell lac is, in fact, prepared by boiling the seed lac with yellow arsenic in a certain proportion. Thus the shell lac is most purified form of lac.

The quality of lac depends upon the host plant. Kusumi lac is said to be the best lac while Dhak is supposed to be the worst and cheapest one. The quality and colour of the lac is variable according to the presence of gum and resins in the host plants.

Composition of Lac: Lac is a complex substance having large amount of resins, together with sugar, water and other alkaline substances. The percentage of various constituents are as given below:

(1) Resin	—	68 to 90%
(2) Dye	—	2 to 10%
(3) Wax	—	6%

(4) Albuminous matter — 5 to 10%

(5) Mineral matter — 3 to 7% and

(6) Water — 3%

Properties of Lac

(1) Lac is not soluble in water but easily soluble in alcohol. This property of lac has great value for insulation of electrical connections.

(2) Lac is easily fusible on heating.

(3) Lac has adhesive quality.

(4) It has binding property when mixed with alcohol.

(5) Lac is also soluble in weak alkali like ammonia.

(6) Lac is a bad conductor of heat.

Enemies of Lac Cultivation: Lac cultivation is destroyed by biotic and abiotic factors:

1. *Abiotic Enemies:* These are high intensity of light, high temperature, high humidity, heavy rainfall and flow of wind.
2. *Biotic Enemies:* The main biotic enemies of lac cultivation are mammals and insects. Krishnaswami *et. al.* (1957, 59), and Gepulpure et. al. (1963), have reported that squirrel, rats, and monkeys cause great damage to the lac crop.

The insects are very powerful enemies of lac crop. Annual loss due to the insect enemies is to the tune of about 4 lakh maunds. The insects damage the crops in different ways.

Parasites: The lac insects are parasitised by eight species of chalcidoid parasites like, *Parenchthrodryinus clavicornis, Erencyrtus dewitzii, Tachardiaephagus tachardiae, Tachardiaephagus tachardiae* var. *somervilli, Eupelmus tachardiae, Coccophagustschirchii, Mariettajavensis* and *Tetrastichus purpureus.* These parasites lay their eggs into lac insects and parasitised

4.8 to 9.9% of lac insects per year and 1/3 of the parasitised cells are males. Thus, it may be concluded that parasitisation is not a major cause of the damage to the lac cultivation.

Predators: Predators cause very severe damage to lac cultivation and 35% of the lac cells are damaged by two predators, *viz., Eublemma amabilis* Moore (Lepidoptera: Noctuidae) and *Holocera pulverea* Meyr (Lepidoptera: Blastobasidae). Female lays eggs near encrusted twigs from where larva emerges and feeds on lac insects.

The Awareness

(1) Twigs for inoculation should be cut just before the swarming to get healthy brood.

(2) Twigs used for inoculation should be free from predators and parasites.

(3) Twigs tied for inoculation should be removed from inoculated host plants after a maximum period of 20 days.

(4) Lac left on the host tree for swarming should be removed in October and November.

(5) The brood lac after swarming should be destroyed along with predators and parasites on it.

(6) The lac scraped from the tree should be taken away from the area of lac infected trees.

(7) Fumigation and water immersion of lac, before removing from twig, should be done.

Lac Industry: India used to produce about 97 per cent of the total lac output in the world but at present it has come down to 50-60 per cent. The cultivation of lac has been good source as an earner of foreign currency. About 50 per cent of the total lac produced in India is obtained from Chhotanagpur area. States like Orissa, Punjab, Madhya Pradesh, W. Bengal, Uttar Pradesh, Gujarat, Rajasthan, Assam, etc. are increasing the production of lac Nowadays. On a very small level lac producing is also reported from Delhi and Kashmir. The average yearly yield of lac in India is about 15,000 metric tons. A lac research institute 'Indian Lac Research Institute' Namkum, Ranchi had been established in 1925 which is producing good quality of white lac. The Indian white lac is supposed to be better than red or other coloured lac because they produce stain or spots at places where they are kept. This is mostly small scale industry with around 350 factories, mostly located in Bihar. In Mirzapur district alone there are about 40 factories. Out of total lac produced in India about 85 to 95 per cent is exported specially to Britain, U.S.A., Russia and West Germany.

Practical Aspects

In 19th century lac dye was in more use than lac resin. Presently due to availability of a better and cheaper annaline dyes the use of lac as a dye has been discarded. The manifest uses of lac is one of the Nature's standing gifts. The various used to which it is put are:

(1) It is utilised in the preparation of gramophone records. Previously this industry utilised major part of the lac produced annually. But nowadays to a great extent plastic is being used in this trade.

(2) It is of utility to Jewellers and Goldsmiths who use lac a filling material in the hollows in gold ornaments like bracelets, armlets and necklaces, etc.

(3) It is an essential gradient used extensively for making polishes, paints and varnishes for finishing wooden as well as metal furnitures and doors, etc.

(4) It is utilised for the preparation of toys, buttons, in pottery and artificial leather.

(5) It is used in the manufacture of photographic material, lithographic ink and for stiffening felt and hat materials.

(6) It is used as an insulating material for electrical goods.

(7) It is also used in confectionery trade as antifowling for applying on ship bottom, grinding stone industry and for ammunition and fire works.

(8) Last but not the least used commonly as sealing wax.

Thus, it is of great use and considered to be as one of the cash crops for the cultivators and also to the Government as source of foreign exchange earners which amount to crores of rupees annually.

Pest Management

The Entomology consist from "Greek" words. The means of "Entomo" is insect and means of "Logos" is knowledge.

- Entomology is the branch of zoology that deals with the study of insects under class insecta of phylum "Arthropods".
- The lac insect *Laccifer lacca* Kerr belongs to the order Hemiptera.
- The silk worms belong to the insect order Lepidoptera.
- Honeybees belong to the order of "Hymenoptera".
- It has been estimated that 50-70% of the grain crops are pollinated by insects, a major portion of it by bees alone.
- The latest production of honey in India is estimated at about 25000-30000 tonnes, its major portion obtained from "Apis *dorsata*".
- The all insects deficit RBC in blood.
- The blood colour of insects is green with yellow.
- The latest estimated total annual losses caused by insect pests to major crops and the food-grains in storage in India to the Rs. 336.6 billion.

- The world wide pre-harvest and post-harvest food losses due to pests have been estimated to be about 48%.
- There is a higher percentage of 46% loss for rice and a lower percentage of 24% for wheat.
- The higher loses occur in Africa 42% and Asia 43% respectively of their potential crop value.
- By the controlling insect-pests of sugarcane the yield could be increased by 20% from 229.2 to 284.9 million tonnes leading to an additional income of 43730 millions of rupees.
- Plant genetic system, a Belgium Biotechnology company in July 1987 were the first reported the development of transgenic plants of tobacco.
- Over 2.5 billion kg of pesticides are applied in the world annually.

Table: Pesticides Consumption durint 1999-2000 in India

Insecticides	60%
Fungicides	21%
Herbicides	14%
Others	5%

- Punjab have the major consumption technical grade pesticide in India, approximately 7150 metric tonnes.
- Lowest consumption of pesticide in India is Daman and Diu, approximately 1 metric tonnes.
- Bartlett (1956) coined the term "Integrated Pest Control"

The term "pest management" was given by Geier (1970).

Control of Integrated Pest: IPM as a dynamic and constantly evolving approach to crop protection in which all the suitable management tactics and available surveillance and forecasting information are utilised to develop a holistic management programme as part of a sustainable crop production technology (By Dhaliwal and Arora 2001)

- Destructive Insect and Pest Act was enforced in the year of 1971.
- The maximum percentage of insecticide is used in cotton in India.
- Average damage of food grains during stored period in India is 10%.
- Plant Protection and Quarantine Act was passed in the year of 1914.
- First insecticide discovered was D.D.T.
- The maximum amount of sugar in honey is 73%.
- 68% resin found in lac insect. *(U.P. Lecturship 2003).*
- Central Honey Bee Research Station is situated at Pune (Maharashtra).
- Lac Research Station is situated at (Bihar).
- Central Silk Research Station is situated at West Bengal.
- D.D.T. was first of all made by "Zedler" scientist.
- The first synthetic pyrethroid was described in 1973.
- LD_{50}: Lethal dose required for killing 50% population of test animals is known as LD_{50} value. The unit of LD_{50} is mg/kg of body weight of test animals or Hg/g body weight of insect.
- Biology has two subdivisions. BOTANY-which deals with plants and ZOOLOGY-which deals with animals.
- The aim of zoology is to collect knowledge of animals.

Branches of Zoology

Entomology — Study of Insects
Protozoology — Study of Protozoa
Nematology — Study of Nematodes
Rodentology — Study of Rodents
Ornithology — Study of Birds
Mammology — Study of Mammals
Ichthyology — Study of Fishes
Helminthology — Study of Worms
Acarology — Study of Mites

Branches of Entomology

Insect Taxonomy
Insect Morphology
Insect Anatomy
Insect Histology
Insect Physiology
Insect Toxicology
Insect Pathology Insect Parasitology
Insect Genetics

Entomology in Application

Agricultural Entomology
Medical Entomology
Horticultural Entomology
Veterinary Entomology
Storage Entomology
Forest Entomology
Industrial Entomology

Insects and their Structure

The Rasping-sucking mouth parts are intermediate in structure between the piercing-sucking and chewing type and are found in Thrips.

The Chewing-lapping type of mouthparts are found in honeybees.

The Piercing-sucking type of mouthparts are found in bugs and mosquitoes.

The Sponging-type of mouth parts are common in flies the typical example being that of the housefly.

- *Pest:* A pest is an organism whose population often increase above a certain level of economic injury and its existance conflicts with man's welfare, convenience and profits.

Sorts of Pests: There are three categories of pests

(i) *Regular Pests:* These are generally found in abundance during a crop season e.g. aphids, Jassids, and Thrips.

(ii) *Sporadic Pests:* These assume pest status occasionally in certain years and include locusts, grasshoppers, hairy caterpillars, crickets and cutworms.

(iii) *Potential Pests:* These pests normally cause negligible damage but may become highly destructive resulting from some disturbance in the environment and the consequent increase in their number e.g., armyworm on wheat.

Step to Control Insect-Pests

(a) *Legislative Measures:* Legislative control involves enactment of laws to regulate the entry, establishment and spread of pests.

Legislation is also imperative to stop the accidental entry, from outside the country of certain pests which may not be present in that country.

Thus legislation is of four kinds

(i) Legislation for foreign quarantine to prevent the introduction of new pests from abroad.

(ii) Legislation for domestic quarantine to prevent the spread of established pests within the country or within a particular state.

(iii) Legislation for notified campaign of control against pests.

(iv) Legislation to prevent the adulteration and mishandling of insecticides or other devices used for the control of pests.

(b) Cultural Practices: The cultural methods of insect control comprise regular form operations, which are so performed as to destroy the insects or to prevent them from causing injury.

For achieving cultural control, the various agricultural practices can be categorised under various groups

1. Tillage
2. Clean seed
3. Irrigation
4. Fertilizers
5. Clean culture
6. Crop rotation
7. Trap crops
8. Pruning and Thining
9. Intercropping
10. Destruction of crop residues.

(c) Mechanical and Physical Control: Mechanical and physical control measures involve the use of force or physical factors of the environment with or without the aid of special equipment.

Hands at Work

It means working with hands. Sometimes with the aid of some simple equipment, like bags, nets, etc.

(d) Chemical Control: The use of chemicals for the control of insect-pests. The classification of insecticides is as follows.

Classification of Insecticides

Insecticides can be grouped in various ways:

According to the Mode of their Entry

(1) Stomach Poisons: They are generally used against insects with chewing type of mouthparts and under certain conditions against those with sponging, siphoning, lapping or sucking mouth parts.

(2) Systemic Poisons: A systemic insecticides when applied to seeds, roots, stems or leaves of plants is absorbed and translocated to various parts of the plant in amounts lethal to insects which feed on them, like-dimethoate, phosphamidon, phorate, aldicarb, etc.

(3) Contact Poisons: The contact poisons are applied as sprays or dusts either directly onto the body of insects or to the places frequented by them. These poisons kill the insects either by clogging spiracles and respiratory system or by entering through the cuticle into the blood and acting as nerve or general tissue poisons. like-chlorinated hydrocarbons (DDT, HCH, aldrin, etc.) carbamates (carbaryl, aldicarb, carbofuran, etc.) and organophosphates (Malathion, Monocrotophos, etc.)

(4) Fumigants: Poisonous gases, derived from either solids or liquids are used as fumigants to kill insect pests of stored grains and other products in warehouses, museum, godowns, etc.

- The common fumigants are-sulphurdioxide, phosphine, methylbromide, EDB, etc.

According to Mode of Action

Physiological poisons

Protoplasmic poisons

Respiratory poisons

Nerve Poisons

According to their Chemical Composition

(i) Elements: Such as sulphur, phosphorus, thallium, mercury, etc.

(ii) Inorganic Compounds: Such as lead arsenate, sodium fluosilicate, zinc phosphide, paris green, etc.

(iii) Organic compounds

(a) Compounds of plant origin such as pyrethrum, nicotine, rotenone, etc.

(b) Animal and mineral oils such as fish oil, diesel oil, etc.

(c) Synthetic organic compounds such as DDT, Malathion, carbaryl, etc.

(iv) Poisonous Gases: Such as hydrogen cyanide ethylene dichloride, carbon tetrachloride, methyl bromide, phosphine, etc. used as fumigants.

(a) Nematicides: These chemicals are used for controlling plant parasitic nematodes. Some of common nematicides are mentioned here.

(1) DBCP (2) D-D Mixture

(3) Ethylene dibromide (4) Ethoprop

(5) Phorate (6) Aldicarb

(b) Rodenticides: These chemicals are used for control of rodents such as-strychnine, warfarin, zinc phosphide, etc.

(c) Molluscicides: Molluscisides may be classified into two broad categories.

(i) Aquatic Molluscicides: copper sulphate niclosamide, sodium pentachlorophenate, etc.

(ii) Terrestrial Molluscicides: include carbamates-aminocarb, methiocarb, mexacarbate, etc.

(d) Fungicides: These chemicals used against fungus like

(i) Bordeaux mixture

(ii) Sulphur and allied compounds

(iii) Carbamates

(iv) Organic mercury compounds

(v) Antibiotics

(vi) Nitrogen compounds Pesticide Application Equipment

The various equipment used include dusters, sprayers, vapour and smoke generators, agricultural aircrafts, granule applicators, etc.

The Insecticide Dusters: The insecticides dust which are very fine particles when falling free in air either slowly or driff for a long distance due to wind.

- Parts of the Dusters

(i) Hopper or Container: It is used for holding the dust.

(ii) Blower: It work to throw the dust outside from container to the discharge outlet.

(iii) Handle: It is used to rotate the gear and there by fan throws the air from blower.

(iv) Feeding Brush: It agitates the dust in the hopper to prevent uneven dusting.

(v) Feed Mechanism: It control the rate of powder coming from hoppers.

(vi) Discharge Line: It carries the powder from bowler to nozzle.

(vii) Shoulder Streps: Shoulder streps are provided with mounting frame in some dusters to set whole mechanism on the chest.

Kinds of Dusters

(a) Manually Operated Dusters

(i) Plunger duster

(ii) Bellows dusters

(iii) Rotary or fan dusters

(iv) Wet dusting equipment

(b) Power Operated Dusters

(i) Tractor mounted duster

(ii) Engine-operated duster

Merits of Dusters

1. Large area can be covered
2. Less expensive
3. Easy to operate
4. Very effective in arid zone area Demerits
5. Loss of chemicals due to drift problem
6. Less efficient
7. Increase toxic hazzard
8. Sprayers

A sprayer is an appliance which atomises the spray fluid, which may be a suspension, an emulsion or a solution, the fluid is ejected with some force for proper distribution.

The various types of sprayers can be grouped into four categories, depending upon the volume of spray fluid discharge to cover a unit area.

1. High Volume Sprayer: These require 300-500 litres per ha. of spray fluid. For example-knapsack sprayer, stirrup sprayer, rocker sprayer, compression sprayer, foot sprayer, etc.
2. Low Volume Sprayer: These sprayers require 50-100 litres of spray fluid per ha. for instance-motorised knapsack sprayers.
3. Ultra Low Volume: These sprayers require 1-5 litres per ha. of spray fluid. For example knapsack mist blowen fogair.
4. Aerosols: Less than 1.0 litres per ha. of spray fluid is required to create fog. The example are-pressurised containers, swing fog machine.

Sprayers and Kinds

(a) Manually Operated Sprayers

These are of two types:

(i) Pneumatic sprayers:

- (a) Hand compression sprayers
- (b) Pressure retaining pumps.
- (c) Small pneumatic atomisers.

(ii) Hydraulic sprayers:

- (a) Knapsack sprayer (b) Foot sprayer
- (c) Rocker sprayer (d) Stirrup sprayer

(b) Power Operated Sprayers

(i) Mist blower or motorised knapsack sprayer

(ii) Tractor mounted sprayer

(iii) Power sprayer

(iv) Ultra-low-volume sprayer

(v) Aerosol dispensers

(vi) Smoke generator

(vii) Vapour generators.

(c) Agricultural Aircrafts

(i) Light aircraft

(ii) Medium aircraft

(iii) Heavy aircraft

(d) Granule Applicator

Placement of insecticidal granules in the whorls of crops like-maize, jawar, etc., is very effective for the control of stem borers.

(e) Soil Injectors

Most soil fumigants are liquids and are applied with soil injectors.

- Merits of sprayers

1. Less drift problems as compared to dusters.
2. It is handy and simple to use.
3. Used to apply various types of chemical such as bactricide, fungicide, insecticide, etc.
4. Most commonly used appliances.
5. Covers more area in less time.
6. Treatment can be done on height and also blow the leaves.

Sustainable pest management is otherwise known as IPM, i.e., Integration of tactics of control a single pest on one or more crops. IPM as applied in agriculture is ideally the use of most effective, economical, safest, ecologically sustainable and sociologically acceptable combination of physical, chemical and biological methods to limit the harmful effects of crop pests.

According to FAO panel of experts (1966-72) IPM is-"A pest management system that in the context of the associated environment and the population dynamics of the pest species,

utilises all suitable a manner as possible and maintains the pest population at levels below those causing economic injury."

The overall objective of IPM is to create and to maintain situations in which insects are prevented from causing significant damage to crops.

A number of tools or components have been successfully used for IPM in several crops. These include-use of pest resistant or tolerant, predators and pathogens, use of parasites, summer ploughing, quarantine measures, late planting, hard collection and destruction, judicious use of pesticides, attractants, repellents, sterilants, growth regulators.

- The sustainable pest management in India has yet to come to be well recognised, but it is still more of an aspiration than a reality for the average farmers.

Biopesticides and their Significance: Biopesticides are most important, effective and commercially viable as these are inexpensive cause no pollution. Pose no risk to human health and have a long term capability to control pests. Biopesticide agents are abundantly available in nature. Several pathogens including viruses such as-NPV (Nuclear polyhedrosis viruses and granulosiss viruses (G.V.)

- Bacteria like-*Bacillus thuringiensis*.
- Fungi like-Metarhizium, Beauveria, Verticillium.
- Protozoa like-*Schizogregrine* cause diseases in insects and destroy them. Similarly, several insect-parasitoids (Parasites thriving on insects are also known in nature.
- *Trichoderma* is an egg parasitoid of several pests.

	Common Name	Scientific Name	Order	Host Plants	Damaging Stage	Mouth Parts
1.	Locust	*Schistocera gregaria* Forsk	Orthoptera	Polyphagous adult	Nymph and cutting type	Chewing and
2.	Termite	*Odontotermes obesus* Ramb	Isoptera	Polyphagous	Nymph and adult	Chewing and cutting type
3.	White Grub	*Holotrichia consanguinea*	Coleoptera	Polyphagous	Grub and adult	Biting and cutting type
4.	White Fly	*Alerolobus baradensis* Mask	Hemiptera Homoptera	Sugarcane, Jawar, Bajra, Wheat, Barley	Nymph and adult	Piercing and sucking type
5.	Leaf Hopper or Pyrilla	*Pyrilla perpusilla* Walk	Hemiptera Homoptera	-do-	-do-	-do-
6.	Sugarcane top borer	*Tryporyza nivella*	Lepidoptera	Sugarcane, Jawar, Bajra, Kans, Sarkanda	Larva	Biting and cutting type
7.	Shoot borer	*Chilo infuscatellus* Snell	Lipidoptera Jawar, Maize	Sugarcane,	Larva	Biting and cutting type

Contd....

	Common Name	Scientific Name	Order	Host Plants	Damaging Stage	Mouth Parts
8.	Paddy stem borer	*Tryporyza incertulas*	-do-	Paddy	-do-	-do-
9.	Rice gundhi bug	*leptocorisa varicornis*	Hemiptera	Paddy, Jawar, Bajra Maize	Nymph and adult	Piercing and sucking type
10.	Paddy plant hopper	*Nephotettix* spp.	Hemiptera	Paddy	-do-	-do-
11.	Cotton Jassid	*Amrasca bigutulla bigutulla*	Hemiptera	Cotton, bhindi	Nymph and adult	-do-
12.	Cotton leaf roller	*Sylepta derogata*	Lepidoptera	-do-	Larva	Chewing and cutting type
13.	Spotted boll worm	*Earias vitella*	-do-	-do-	-do-	-do-
14.	Pink boll worm	*Pectinophora gossypiella* Sound	Lepidoptera	-do-	-do-	-do-
15.	Mustard aphid	*Lipaphis erysimi*	Hemiptera	Radish, Rai, Mustard, Cauliflower, turnip, etc.	Nymph and adult	Piercing and sucking type

Contd....

	Common Name	Scientific Name	Order	Host Plants	Damaging Stage	Mouth Parts
16.	Painted bug	*Bagrada cruciferorum*	Hemiptera	-do-	-do-	-do-
17.	Cut worm	*Agrotis ypsilon*	Lepidoptera	Mainly gram and potato	Larva	Cutting and biting type
18.	Gram caterpillar	*Helicoverpa armigera*	-do- tomato, Arhar	Gram, pea,	-do-	-do-
19.	Red pumpkin beetle	*Raphidopalpa foveicollis*	Coleoptera	Plants of cucurbitaceae	Grub and adult	-do-
20.	Fruit fly	*Dacus cucurbitae*	Diptera	Plants of cucurbitaceae	Maggot	-do-
21.	Epilachna beetle	*Epilachna vigintioctapun-ctata*	Coleoptera	Brinjal and cucurbits	Grub and adult	-do-
22.	Brinjal shoot and fruit borer	*Leucinodes arbonatis*	Lepidoptera Potato	Brinjal and	Larva	-do-
23.	Diamond back moth	*Plutella maculipensis*	Lepidoptera cruciferae	Plants of biting type	-do-	Cutting and

Contd....

	Common Name	Scientific Name	Order	Host Plants	Damaging Stage	Mouth Parts
24.	Tobacco caterpillar	*Prodenia litura*	Lepidoptera	Tobacco and cruciferae	-do-	-do-
25.	Mango hopper	leaf *Amritkodus atkinsoni*	Hemiptera Homoptera	Mango adult	Nymph and sucking type	Piercing and
26.	Mango mealy bug	*Drosicha mangifera green*	Hemiptera Homoptera	Mango, Jack fruit, etc.	-do-	-do-
27.	Lemon buterfl	*Popilio demoleus* Linn	Lepidoptera	Citrus	Larva	Biting and cutting type
28.	Ber fruit fly	*Carpomyia vesuviana*	Diptera	Ber	Larva	-do-
29.	Rice weevil	*Sitophilus oryzae* Linn.	Coleoptera	Stored grain	Grub and adult	Chewing and cutting type
30.	Khapra beetle	*Trogodemua granarium*	Coleoptera	Stored grain	Grub	-do-
31.	Grain and flour moth	*Siiotroga cerealella.*	Lepidoptera	-do-	Larva	-do-
32.	Pulse beetle	*Callosobruchus chinensis*	Coleoptera	Pulses	Grub and adult	-do-

Contd....

Common Name	Scientific Name	Order	Host Plants	Damaging Stage	Mouth Parts
33. Lesser grain borer	*Rhizopertha dominica*	Coleoptera	Stored grain	-do-	-do-
34. Red hairy caterpiller	*Amsacta spp.*	Lepidoptera	Polyphagous	Larva	Biting and cutting type
35. Gurdaspur borer	*Acigona steniellus*	Lepidoptera	Maize, Jawar, Sugarcane	Larva	-do-
36. Army worm	*Mythimna separata*	Lepidoptera	Paddy, Jawar, Bajra, Wheat, Maize, Sugar.	Larva	Chewing and cutting type
37. Shoot fly	*Atherigona varia soceata*	Diptera	Jawar, Bajra.	Maggot	-do-
38. Fruit sucking	*Ophideres* spp.	Lepidoptera	Mango, Guava, Citrus and Pomegranate	Moth	Siphoning moth type
39. Potato tuber moth	*Gnorinoschema operculella*	Lepidoptera Tomato, Tobacco and	Potato,	Larva	Chewing and cutting type
		Agricultural Entrance		*108*	Brinjal

Important Glossary

- Acaricide or Miticide: The chemical substances used to kill mites are known as acaricide.
- Adulticide: Chemical substance used to kill adult insects.
- Antibiosis: Antibiosis refers to the adverse effect of the host plant on the biology of the insects and their progeny infesting it.
- Allelochemicals: Allelochemicals are non-nutritional chemicals produced by an organism of one species and affect the growth, health, behaviour or population biology of individuals of another species.
- American Bolworm: Recently the American bollworm has become a serious pest of cotton in northern parts of India. It was first reported on cotton in 1977 in Punjab.
- Armyworm: The armyworm is a pest of graminaceous crops-all over the world like jawar, maize, bajra, etc.
- Antifeedants: The antifeedants are chemicals which inhibit or deter the feeding of insects due to their presence on the natural food of the species concerned. An antifeedant acts by suppressing the gustatory receptors.
- Avermectins: Avermectins are macrocyclic lactones which were originally isolated in 1976 by scientists at Merck and Co, Inc from a culture of *Streptomyces avermitilis* from Japan.
- Allomone: A compound released by one organism which evokes a reaction in an individual of a various species that is favourable to the emitter but not to the receiver.
- Antimone: A substance produced or acquired by an organism that when it contacts an individual of another species in the natural context, evokes in the receiver a behavioural or physiological reaction that is maladaptive to both the emitter and the receiver.

4

Resources of Earth

Land as a Resource: Land is a finite and valuable resource upon which we depend for our food, fibre and fuel wood, the basic amenities of life. Soil, especially the top soil, is classified as a renewable resource because it is continuously regenerated by natural process though at a very slow rate. About 200-1000 years are needed for the formation of one inch or 2.5 cm soil, depending upon the climate and the soil type. But, when rate of erosion is faster than rate of renewal, then the soil becomes a non-renewable resource.

Land Degradation: With increasing population growth the demands for arable land for producing food, fibre and fuel wood is also increasing. Hence there is more and more pressure on the limited land resources which are getting degraded due to over-exploitation. Soil degradation is a real cause of alarm because soil formation is an extremely slow process as discussed above and the average annual erosion rate is 20-100 times more than the renewal rate.

Soil erosion, water-logging, salinisation and contamination of the soil with industrial wastes like fly-ash, press-mud or heavy metals all cause degradation of land.

Soil Erosion: The literal meaning of 'soil erosion' is wearing away of soil. Soil erosion is defined as the movement of soil components, especially surface-litter and top soil from one place to another. Soil erosion results in the loss of fertility because it is the top soil layer which is fertile. If we look at the world situation, we find that one-third of the world's cropland is getting eroded. Two-third of the seriously degraded lands lie in Asia and Africa.

Soil erosion is basically of two types based upon the cause of erosion:

(i) *Normal Erosion or Geologic Erosion:* caused by the gradual removal of top soil by natural processes which bring an equilibrium between physical, biological and hydrological activities and maintain a natural balance between erosion and renewal.

(ii) *Erosion at Accelerated Pace:* This is mainly caused by anthropogenic (man-made) activities and the rate of erosion is much faster than the rate of formation of soil. Overgrazing, deforestation and mining are some important activities causing accelerated erosion. There are two types of agents which cause soil erosion:

(a) *Climatic Agents:* water and wind are the climatic agents of soil erosion. Water affects soil erosion in the form of torrential rains, rapid flow of water along slopes, runoff, wave action and melting and movement of snow.

Water induced soil erosion is of the following types:

- *Sheet Erosion:* when there is uniform removal of a thin layer of soil from a large surface area, it is called sheet erosion. This is usually due to runoff water.
- *Rill Erosion:* When there is rainfall and rapidly running water produces finger-shaped grooves or rills over the area, it is called rill erosion.

- *Gully Erosion:* It is a more prominent type of soil erosion. When the rainfall is very heavy, deeper cavities or gullies are formed, which may be U or V shaped.
- *Slip Erosion:* This occurs due to heavy rainfall on slopes of hills and mountains.
- *Stream Bank Erosion:* During the rainy season, when fast running streams take a turn in some other direction, they cut the soil and make caves in the banks.

  Wind erosion is responsible for the following three types of soil movements:
- *Saltation:* This occurs under the influence of direct pressure of stormy wind and the soil particles of 1-1.5 mm diameter move up in vertical direction.
- *Suspension:* Here fine soil particles (less than 1 mm diameter) which are suspended in the air are kicked up and taken away to distant places.
- *Surface Creep:* Here larger particles (5-10 mm diameter) creep over the soil surface along with wind.

  (b) *Biotic Agents:* Excessive grazing, mining and deforestation are the major biotic agents responsible for soil erosion. Due to these processes the top soil is disturbed or rendered devoid of vegetation cover. So the land is directly exposed to the action of various physical forces facilitating erosion. Overgrazing accounts for 35% of the world's soil erosion while deforestation is responsible for 30% of the earth's seriously eroded lands. Unsustainable methods of farming cause 28% of soil erosion.

Deforestation without reforestation, overgrazing by cattle, surface mining without land reclamation, irrigation techniques that lead to salt build-up, water-logged soil, farming on land with unsuitable terrain, soil compaction by agricultural machinery, action of cattle trampling, etc. make the top soil vulnerable to erosion.

Preserving Soil

In order to prevent soil erosion and conserve the soil the following conservation practices are employed:

(i) *Conservational till Farming:* In traditional method the land is ploughed and the soil is broken up and smoothed to make a planting surface. However, this disturbs the soil and makes it susceptible to erosion when fallow (i.e. without crop cover). Conservational till farming, popularly known as no-till-farming causes minimum disturbance to the top soil. Here special tillers break up and loosen the subsurface soil without turning over the topsoil. The tilling machines make slits in the unploughed soil and inject seeds, fertilizers, herbicides and a little water in the slit, so that the seed germinates and the crop grows successfully without competition with weeds.

(ii) *Contour Farming:* On gentle slopes, crops are grown in rows across, rather than up and down, a practice known as contour farming. Each row planted horizontally along the slope of the land acts as a small dam to help hold soil and slow down loss of soil through runoff water.

(iii) *Terracing:* It is used on still steeper slopes are converted into a series of broad terraces which run across the contour. Terracing retains water for crops at all levels and cuts down soil erosion by controlling run off. In high rainfall areas, ditches are also provided behind the terrace to permit adequate drainage.

(iv) *Strip Cropping:* Here strips of crops are alternated with strips of soil saving covercrops like grasses or grass-legume mixture. Whatever runoff comes from the cropped soil is retained by the strip of cover-crop and this reduces soil erosion. Nitrogen fixing legumes also help in restoring soil fertility.

(v) *Alley Cropping:* It is a form of inter-cropping in which crops are planted between rows of trees or shrubs. This is also called agro-forestry. Even when the crop is harvested, the soil is not fallow because trees and shrubs still remain on the soil holding the soil particles and prevent soil erosion.

Wind breaks or shelterbelts: They help in reducing erosion caused by strong winds. The trees are planted in long rows along the cultivated land boundary so that wind is blocked. The wind speed is substantially reduced which helps in preventing wind erosion of soil.

Thus, soil erosion is one of the world's most critical problems and, if not slowed, will seriously reduce agricultural and forestry production, and degrade the quality of aquatic ecosystems as well due to increased siltation. Soil erosion, is in fact, a gradual process and very often the cumulative effects becomes visible only when the damage has already become irreversible. The best way to control soil erosion is to maintain adequate vegetational cover over the soil.

Water-logging: In order to provide congenial moisture to the growing crops, farmers usually apply heavy irrigation to their farmland. Also, in order to leach down the salts deeper into the soil, the farmer provides more irrigation water. However, due to inadequate drainage and poor quality irrigation water there is accumulation of water underground and gradually it forms a continuous column with the water table.

We call these soils as waterlogged soils which affect crop growth due to inhibition of exchange of gases. The pore-spaces between the soil particles get fully drenched with water through the roots.

Water-logging is most often associated with salinity because the water used for irrigation contains salts and the soils get badly degraded due to erroneous irrigation practices. The damages caused by some major irrigation projects is shown is below.

Water-logging and Salinisation caused due to Some Irrigation Projects in India

Irrigation Project	*State*	*Area affected (thousand hectares)*	
		Water logging	*Salinity*
Indira Gandhi Canal	Rajasthan	43	29
Gandak	Bihar, Gujarat	211	400
Chambal	M.P., Rajasthan	98	40
Ram Ganga	UP.	195	352
Sri Ram Sagar	Andhra Pradesh	60	1

Source: B.K. Garg and I.C. Gupta (1997).

An estimated loss of Rs. 10,000 million per annum occurs due to water-logging and salinity in India. It is a startling fact because the cost of development of the irrigation projects is very high and in the long run they cause problems like water-logging and salinity thereby sharply reducing soil fertility.

Landslides: Various anthropogenic activities like hydroelectric projects, large dams, reservoirs, construction of roads and railway lines, construction of buildings, mining, etc. are responsible for clearing of large forested areas. Earlier there were few reports of landslides between Rishikesh and Byasi on Badrinath Highway area. But, after the highway was constructed, 15 landslides occurred in a single year. During construction of roads, mining activities, etc. huge portions of fragile mountainous areas are cut or destroyed by dynamite and thrown into adjacent valleys and streams. These land masses weaken the already fragile mountain slopes and lead to landslides. They also increase the turbidity of various nearby streams, thereby reducing their productivity.

Desertification: Desertification is a process whereby the productive potential of arid or semiarid lands falls by ten per cent or more. Moderate desertification is 10-25% drop in productivity, severe desertification causes 25-50% drop while very severe desertification results in more than 50% drop in productivity and usually creates huge gullies and sand dunes. Desertification leads to the conversion of rangelands or irrigated croplands to desert-like conditions in which agricultural productivity falls.

Desertification is characterised by de-vegetation and loss of vegetable cover, depletion of groundwater, salinisation and severe soil erosion. Desertification is not the literal invasion of desert into a non-desert area. It includes degradation of the ecosystems within as well as outside the natural deserts. The Sonoran and Chihuahuan deserts are about a million years old, yet they have become more barren during the last 100 years. So, further desertification has taken place within the desert.

Causes of Desertification: Formation of deserts may take place due to natural phenomena like climate change or may be due to abusive use of land. Even the climate change is linked in many ways to human activities. The major anthropogenic activities responsible for desertification are as follows:

(a) *Deforestation:* The process of denuding and degrading a forested land initiates a desert producing cycle that feeds on itself. Since there is no vegetation to hold back the surface runoff, water drains off quickly before it can soak into the soil to nourish the plants or to replenish the groundwater. This increases soil erosion, loss of fertility and loss of water.

(b) *Overgrazing:* The regions most seriously affected by desertification are the cattle-producing areas of the world. This is because the increasing cattle population heavily graze in grasslands or forests and as a result denude the land area. When the earth is denuded, the microclimate near the ground becomes inhospitable to seed germination. The dry barren land becomes loose and more prone to soil erosion. The top fertile layer is also lost and thus plant growth is badly hampered in such soils. The dry barren land reflects more of the sun's heat, changing wind patterns, driving away moisture-laden clouds leading to further desertification.

(c) *Mining and Quarrying:* These activities are also responsible for loss of vegetal cover and denudation of extensive land areas leading to desertification. Deserts are found to occur in the arid and semi-arid areas of all the Continents. During the last 50 years about 900 million hectares of land have undergone desertification over the world. This problem is especially severe in Sahel region, just south of the Sahara in Africa. It is further estimated that if desertification continues at the present rate, then by 2010, it will affect such lands which are presently occupied by 20% of the human population.

Amongst the most badly affected areas are the sub Saharan Africa, the Middle East, Western Asia, parts of Central and South America, Australia and the Western half of the United States. It is estimated that in the last 50 years, human activities have been responsible for desertification of land area equal to the size of Brazil. The UNEP estimates suggest that if we don't make sincere efforts now then very soon 63% of rangelands, 60% of rain-fed croplands and 30% of irrigated croplands will suffer from desertification on a worldwide scale, adding 60,000 Km^2 of deserts every year.

Natural Resources' Conservation Individual's Part: Different natural resources like forests, water, soil, food, mineral and energy resources play a vital role in the development of a nation. However, overuse of these resources in our modern society is resulting in fast depletion of these resources and several related problems. If we want our mankind to flourish there is a strong need to conserve these natural resources.

While conservation efforts are underway at national as well as international level, the individual efforts for conservation of natural resources can go a long way. Environment belongs to each one of us and all of us have a responsibility to contribute towards its conservation and protection. "Small droplets of water together form a big ocean". Similarly, with our small individual efforts we can together help in conserving our natural resources to a large extent. Let us see how can individuals help in conservation of different resources.

Preserving Water

- Don't keep water taps running while brushing, shaving, washing or bathing.
- In washing machines fill the machine only to the level required for your clothes.
- Install water-saving toilets that use not more than 6 liters per flush.
- Check for water leaks in pipes and toilets and repair them promptly. A small pin-hole-sized leak will lead to the wastage of 640 liters of water in a month.
- Reuse the soapy water of washings form clothes for washing off the courtyards, driveways, etc.
- Water the plants in your kitchen-garden and the lawns in the evening when evaporation losses are minimum. Never water the plants in mid-day.
- Use drip irrigation and sprinkling irrigation to improve irrigation efficiency and reduce evaporation.

- Install a small system to capture rain water and collect normally wasted used water from sinks, cloth-washers, bath tubs, etc. which can be used for watering the plants.
- Build rain water harvesting system in your house. Even the President of India is doing this.

Preserving Energy

- Turn off lights, fans and other appliances when not in use.
- Obtain as much heat as possible from natural sources. Dry the clothes in sun instead of drier if it is a sunny day.
- Use solar cooker for cooking your food on sunny days which will be more nutritious and will cut down on your LPG expenses.
- Build your house with provision for sunspace which will keep your house warmer and will provide more light.
- Grow deciduous trees and climbers at proper places outside your home to cut off intense heat of summers and get a cool breeze and shade. This will cut off your electricity charges on coolers and air-conditioners. A big tree is estimated to have a cooling effect equivalent to five air conditioners. The deciduous trees shed their leaves in winter. Therefore they do not put any hindrance to the sunlight and heat.
- Drive less, make fewer trips and use public transportations whenever possible. You can share by joining a car-pool if you regularly have to go to the same place.
- Add more insulation to your house. During winter close the windows at night. During summer close the

windows during days if using an A.C. Otherwise loss of heat would be more, consuming more electricity.

- Instead of using the heat convector more often wear adequate woolens.
- Recycle and reuse glass, metals and paper.
- Try riding bicycle or just walk down small distances instead of using your car or scooter.
- Lower the cooling load on an air-conditioner by increasing the thermostat setting as 3-5 % electricity is saved for every one degree rise in temperature setting.

Saving Earth

- While constructing your house, don't uproot the trees as far as possible. Plant the disturbed areas with a fast-growing native ground cover.
- Grow different types of ornamental plants, herbs and trees in your garden. Grow grass in the open areas which will bind the soil and prevent its erosion.
- Make compost from your kitchen waste and use it for your kitchen-garden or flower-pots.
- Do not irrigate the plants using a strong flow of water, as it would wash off the soil.
- Better use sprinkling irrigation.
- Use green manure and mulch in the garden and kitchen-garden which will protect the soil.
- If you own agricultural fields, do not over-irrigate your fields without proper drainage to prevent water-logging and salinisation.
- Use mixed cropping so that some specific soil nutrients do not get depleted.

Durable Agriculture Incentives

- Do not waste food. Take as much as you can eat.
- Reduce the use of pesticides.
- Fertilize your crop primarily with organic fertilizers.
- Use drip irrigation to water the crops.
- Eat local and seasonal vegetables. This saves lot of energy on transport, storage and preservation.
- Control pests by a combination of cultivation and biological control methods.

Resources use Equitable Term for Sustainable Life-style: There is a big divide in the world as North and South, the more developed countries (MDC's) and less developed countries (LDC's), the haves and the have-nots. The less developed does not mean that they are backward as such, they are culturally very rich or even much more developed, but economically they are less developed. The gap between the two is mainly because of population and resources.

The MDC's have only 22% of world's population, but they use 88% of its natural resources, 73% of its energy and command 85% of its income. In turn, they contribute a very big proportion to its pollution. These countries include USA, Canada, Japan, the CIS, Australia, New Zealand and Western European Countries. The LDC's, on the other hand, have very low or moderate industrial growth, have 78% of the world's population and use about 12% of natural resources and 27% of energy. Their income is merely 15% of global income. The gap between the two is increasing with time due to sharp increase in population in the LDC's. The rich have grown richer while the poor have stayed poor or gone even poorer.

As the rich nations are developing more, they are also leading to more pollution and sustainability of the earth's life support system is under threat. The poor nations, on the other hand, are still struggling hard with their large population and poverty problems. Their share of resources is too little leading to unsustainability.

As the rich nations continue to grow, they will reach a limit. If they have a growth rate of 10 % every year, they will show 1024 times increase in the next 70 years. Will this much of growth be sustainable? The answer is 'No' because many of our earth's resources are limited and even the renewable resources will become unsustainable if their use exceeds their regeneration.

Thus, the solution to this problem is to have more equitable distribution of resources and wealth. We cannot expect the poor countries to stop growth in order to check pollution because development brings employment and the main problem of these countries is to tackle poverty. A global consensus has to be reached for more balanced distribution of the basic resources like safe drinking water, food, fuel, etc. so that the poor in the LDC's are at least able to sustain their life. Unless they are provided with such basic resources, we cannot think of rooting out the problems related to dirty, unhygienic, polluted, disease-infested settlements of these people which contribute to unsustainability.

Thus, the two basic causes of unsustainability are over population in poor countries who have under-consumption of resources and over-consumption of resources by the rich countries, which generate wastes. In order to achieve sustainable life styles it is desirable to achieve a more balanced and equitable distribution of global resources and income to meet everyone's basic needs.

The rich countries will have to lower down their consumption levels while the bare minimum needs of the poor have to be fulfilled by providing them resources. A fairer sharing of resources will narrow down the gap between the rich and the poor and will lead to sustainable development for all and not just for a privileged group.

Biosphere

The part of the earth containing living organisms is known as 'biosphere'. It can also be defined as 'the part of the earth in which the ecosystem exists'. The physical portion of the biosphere consists of following three subdivisions:

(1) Hydrosphere, which includes all liquid components namely that water in oceans, lakes, rivers and on land;

(2) Lithosphere, which comprises the solid components, the rocky substance of the continents and

(3) Atmosphere, the gaseous mantle around the hydrosphere and lithosphere.

Plants require inorganic metabolites from each of the subdivisions of the biosphere. The hydrosphere supplies liquid water; the lithosphere supplies all other minerals; and the atmosphere supplies oxygen, intergang and carbon dioxide. Together, these inorganic materials provide all the chemical elements needed in the construction and maintenance of living matter.

Role of Hydrosphere

Water represents the most abundant mineral of the plant. It covers some 70% of earth's surface entirely, and it is a major constituent of the lithosphere and the atmosphere. Water is also the most abundant component of living matter.

The basic cycle of water in the environment is quite familiar. Solar energy evaporates water from the hydrosphere into the atmosphere. Later, cooling and condensation of the vapours at higher altitudes produce clouds-precipitation as rain or snow then returns the water to the hydrosphere. This is the most massive process of any kind on earth consuming more energy and moving more material than any other process.

Aquatic animals obtain water from their environment and return a part of it throughout excretion during their life time. After death, the remaining water is returned through decay. Terrestrial organisms, i.e. plants and animals are interposed more extensively in the global cycle of water.

These organisms draw water from the reservoirs present in the soil such as rivers, streams, lakes, ponds, etc. Plants and animals move this water through their bodies where some of it is retained. The rest is excreted, partly as liquid water back into the hydrosphere but more particularly as water vapour into the atmosphere. Actually terrestrial organisms accelerate the global water cycle. After the death of terrestrial organisms, any liquid water present in their body returns to the hydrosphere through decay.

Though a prime nutrient, water also influences animals through its effects on climate and weather, both on sea and on land. In the ocean, water warmed in the tropics becomes light and rises to the surface, whereas cold polar water sinks. These up and down displacements of water along with the driving force of the winds result in ocean currents. Such currents influence climatic conditions not only within the sea but also in the air and on land. As a consequence of its thermal properties, water has another climatic effect. Of all liquids, water is one of the slowest to heat or cool and stores a very large amount of thermal energy.

Thus oceans have become huge reservoirs of solar heat. The result is the sea-air chilled at night becomes less cold because of heat radiation from water warmed by day. Conversely, sea-air warmed during the day, becomes less hot because of heat absorption by water cooled at night. Warm and cool inshore winds then produce seasonal summer and winter patterns.

Long-range global climates are influenced by the relative amount of water locked into polar ice. Temperature variations averaging only a few degrees produce over the years by geophysical changes is suffice for the advance or retreat of polar ice. At present the earth is slowly emerging from the last ice age, and as polar ice is now melting, water levels are rising and coast lines are gradually being submerged.

All these changes in the hydrosphere have profound impact on animals. By influencing temperature, humidity, amount of precipitation, winds, waves and indeed the very presence or absence of water in given localities, they play a major role m determining what kinds and amounts of animals can live in such localities.

Role of Lithosphere

Lithosphere not only supplies most mineral nutrients to all animals but it also forms the sand base of soil, specifically, required by terrestrial plants and by numerous subterranean animals.

Like the world's water, the rocky substance of the earth surface moves in a gigantic cycle. But here the rate of circulation is measured in thousands and millions of years. One segment of this global mineral cycle is diastrophism, the vertical uplifting of large tracts of earth's crust through a variety of geological forces. Major parts of continents or indeed whole continents may undergo such slow diastrophic movements.

The most striking example is mountain building. Presently the youngest and highest mountains are the Himalayas, the Andes and the Alps. All of them were thrown up during the laramid revolution, some 70 million years ago. The earth surface in this region is not completely settled even now.

The appearance of a high mountain barrier is likely to interfere drastically with continental air circulations. The moisture laden oceanwinds may no longer be able to pass across the barrier. Rain will continue to fall in the region between ocean and mountain, which will become lush and fertile. But the region beyond the mountain will be arid and desert conditions are likely to develop. Second segment of the global lithospheric cycle involves gradation, the lowering of high land and the levelling of mountains. These changes result in part from erosion and dissolution of rocks and in part from actual geologic sinking of land by the hydrosphere and the atmosphere.

The physical forces of gradations are chemical forces, which are particularly important to animals. The chemical processes of decay are major erosive factors. They contribute to the breaking of large stones into smaller ones and then into tiny sand grains and microscopic rock fragments. Chemical erosion thus play a major role in the formation of soil particles.

Second, whenever water comes in contact with a rock it dissolves small quantities of it and so acquires any mineral content. Dissolved soil minerals eventually drain into rivers and rivers drain into the ocean. Hence as the lithosphere is slowly being denuded of mineral compounds, the hydrosphere fills with them. It was partly because of it that the early seas on earth acquired their original saltness. Plants in the sea freely use the minerals as metabolites.

The death of marine organisms helps to complete the global mineral cycle. Many protists and animals use mineral nutrients in the construction of protective shells and supporting bones. After these organisms die, their mineral components rain down to the sea floor where, as pointed out earlier, the accumulating layers of ooze may ultimately compress into rock.

The global lithospheric cycle becomes complete when such a section of sea bottom or low lying land is subjected to new diastrophic forces. High ground or mountains are thereby regenerated and parts that were seafloor originally may be thrust up as new land in the process.

The lithospheric cycle supplies many more types of minerals than normally plants require. The minerals required and used by the plants are nitrates, phosphates, chlorides, carbonates, sulphates, and sodium, potassium, calcium, manganese, magnesium, copper and iron. Despite the dense cover of plant life carpeting the earth, most of the minerals are still well in excess of currently required amounts. However, because of the uninterrupted global growth of plants for million of years, a few key minerals now tend to be in relatively short supply. In particular phosphates and nitrates have become significant limiting factors. Thus agriculture soils are artificially enriched with fertilizers, largely phosphates and nitrated.

Role of Atmosphere

Atmosphere is also subjected to physical cycles by the sun and the spin of the earth. Warm equatorial air rises, cool polar air sinks and the axial rotation of earth shifts air mass laterally. Like the water circulation, global air circulation also influences climatic conditions substantially. Equally significant to animals are the chemical cycles of the atmosphere. Air consists mainly of O_2 (20%), CO_2 (0.3%), and N_2 (79%). Water and minute traces of inert gases (He, Ne, Kr and Ar) are also present in the atmosphere. Except inert gases all the components of air serve as raw materials and each plays a conspicuous role in the global cycle.

Complex Cycles

Unlike energy, the chemical elements tend to circulate in the biosphere in characteristic paths from environment to organisms and back to the environment. Since organic substances are in limited supply in available form, their cyclic use and reuse constitutes an important part of the ecosystems.

These more or less circular paths are known as biogeochemical cycles. Thus biogeochemical cycle may be defined as 'the more or less circular path which brings about the circulation of chemical elements, including all essential elements from the environment to the organisms and back to the environment.

Types of Biogeochemical Cycles: The biogeochemical cycles are of two types:

(1) The gaseous cycles in which the reservoir is in the atmosphere or hydrosphere and

(2) The mineral and sedimentary cycles, in which the reservoir is in the lithosphere, i.e. earth's crust.

Cycles of Gases

***Oxygen Cycle*:** Atmospheric oxygen enters in the living world as a respiratory gas. The function of oxygen in respiration is to combine with hydrogen. Thes resulting water becomes part of the general water content of the living matter, as a source of element-hydrogen and oxygen. The structural oxygen remains in the organisms until decay and after death returns to the environment.

However, it usually returns not in the form of free atmospheric oxygen but as either water or carbon dioxide. The global oxygen cycle is interlinked closely with the global water and CO_2 cycles. A third possible fate of water within organisms (green plants) is its use as a raw material in photosynthesis, where it is split apart into hydrogen and oxygen. The hydrogen participates in food manufacture and oxygen as a by product comes out m the atmosphere.

Atmospheric oxygen is also the source of ozone (O_3) layer which envelops the earth at an altitude of some 10 miles. This layer prevents a great deal of the high-energy relation of the sun from reaching the earth's surface and so it affects plants metabolism indirectly by shedding potential lethal rays.

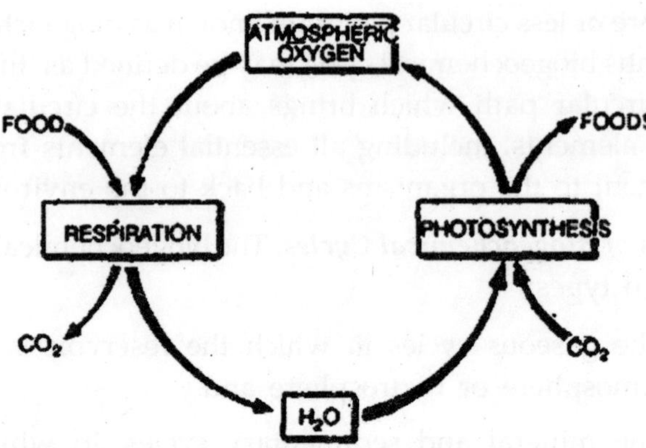

The Oxygen cycle.

Cycle of CO_2

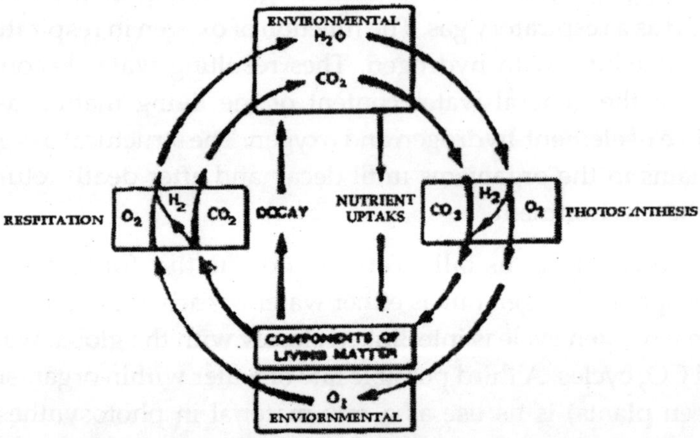

The inter-relation of the oxygen, water and carbon cycles.

Atmospheric carbon dioxide is virtually the exclusive carbon source and with water one of the two major oxygen sources for the construction of living matter. The gas enters the living world through photosynthesis, in which it is a fundamental raw material. Photosynthesis incorporates CO_2 into organic substances and these are partly used in the construction of more living matter.

The CO_2 content of the atmosphere is replenished not only through biological combustion or respiration, but also through non-living combustion of real fires. In the first case CO_2 is a by-product and is returned to the environment immediately; in the second it is a decay product returned after death. The release of CO_2 into the air in forest fibres and in the burning of industrial fuels actually represents a long delayed completion of carbon cycle. The combustible substances in wood, coal, oil and natural gas, all are organic compounds that are manufactured through photosynthesis.

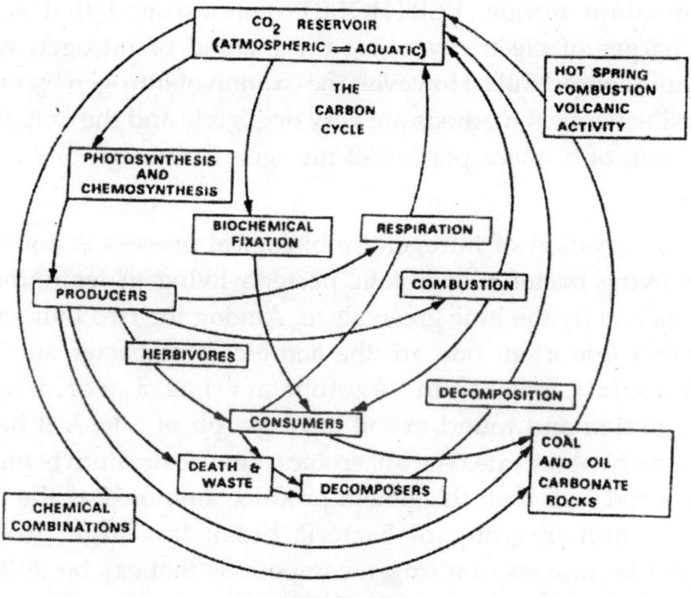

Carbon cycle.

Cycle of Nitrogen: Protein is the main part of protoplasm and nitrogen is an essential constituent of protein. Animals for their very existence and growth require large quantities of nitrogen. Though the atmosphere contains about 79% of nitrogen in the gaseous state, yet animals cannot utilise it unless it is converted into some usable form. Animals take in nitrogen either in inorganic form as ammonia, nitrates and nitrites or in organic form as urea, proteins and nucleic acids. Thus atmospheric nitrogen can be used by animals only after it has been fixed as one of the inorganic substances. This fixation takes place both by biological as well as physio-chemical processes.

It was CAVENDISH who first noticed the increase in the quantities of nitrate in the soil after rain and thundershowers and explained, that during lightning union of nitrogen and oxygen occurs. SCHONBEIN explained that small amounts of nitrogen combine with hydrogen during evaporation to produce ammonium nitrate. BERTHOLET demonstrated that small discharges of electricity cause the fixation of nitrogen with organic compounds. However, the fixation of nitrogen by these physio-chemical methods are very negligible and the only way of fixing of a major portion of nitrogen is through biological process.

The fixation of nitrogen by biological process is done by free living bacteria, symbiotic bacteria living in leguminous plants and by the blue green algae. Among the free living soil bacteria important one are the aerobic Azotobacter and the anaerobic Clostridium. Azotobacter has a world-wide distribution and found in soil having a ph of 6 to 7. It has a high respiratory rate. The anaerobic form Clostridium is found in neutral soil. Both the genera produce ammonia as the first step. Another group of bacteria break the large protein molecules into small nitrogen compounds that can be utilised by roots. It is these bacteria that help in the recovery of nitrogen compounds that are locked in the tissues of plants and animals in the normal process. This breakdown takes place in two parts:

(a) *Ammonification:* In this process ammonia is produced from the organic material of plants and animals. The bacteria that are involved in the process are Bacillus ramosus, B. mycoides and B. proteusvulgaris.

(b) *Nitrification:* Certain other bacteria transform ammonia into nitrates. This takes place in two steps. Nitrosomonas and Nitrococcus bacteria oxidise ammonia to nitrites. These nitrites are again oxidised by the bacteria Nitrobacter to nitrates. The action of these nitrifying bacteria are influenced by environmental and chemical nature of the soil. Both the types of bacteria work well in an alkaline medium.

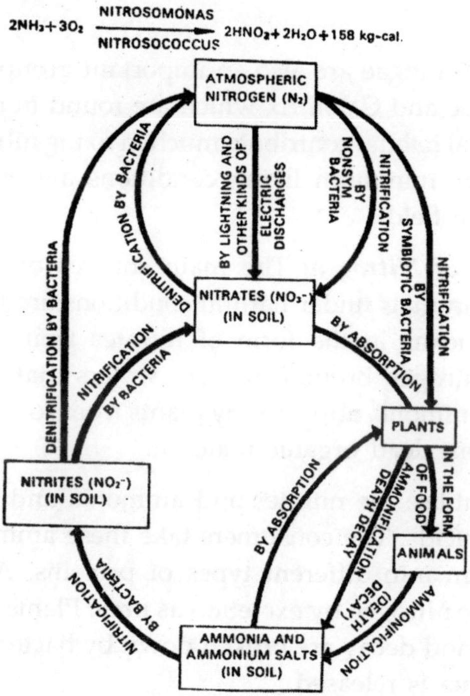

The Nitrogen cycle

The symbiotic bacteria Rhizobium radicicolum lives on the roots of the leguminous plants. The secretion of the plant roots help in multiply the bacteria. The bacterial excretions dissolve the cellulose of the root hairs, move in and bring about the rapid multiplication of cell. It is these cells that form the central mass of the root nodules.

The bacteria now change their shape from rod shape to bacteriod condition and start fixing nitrogen. The leguminous plant's thus fix nitrogen with the help of bacteria in agricultural land.

There are also non-leguminous nodule bearing plants which also do a similar job. They increase the nitrogen content of wide lands. Important non-leguminous plants which do nitrogen fixation are Alnus, Myrica, Ceanothurs and Shepherdia.

Blue green algae are also an important group of nitrogen fixers, Nostoc and Calothrix which are found in both aquatic and terrestrial habital contribute much in fixing nitrogen. These algae require minimum living conditions and are found to thrive in rice fields.

Cycling of Nitrogen: The main sources of nitrogen for plants and animals under natural conditions are that fixed by biological means in the form of nitrates from atmosphere, inorganic nitrogen brought by rain waters that are fixed by lightening ammonia absorbed by plants from soil and nitrogen released from dead organic material.

Plants utilise the nitrates and ammonia and covert them into amino acids. The consumers take these amino acids and convert them into different types of proteins. Animals also release some nitrogen by excretion as urea. Plants and animals after death and decay are broken down by bacteria and fungi and ammonia is released.

Ammonia can be utilised by plants to form nitrates and nitrites or can find their way to the atmosphere. Nitrates and nitrites can be carried away by rivers to lakes and also to the sea. There are certain bacteria called the denitrifying bacteria. Important among them is Bacterium denitrificans. They use nitrates as nutrients and ultimately convert them to molecular nitrogen. As this gas escapes into the atmosphere global cycle becomes complete.The nitrogen cycle is also affected by human intrusion. Grassland cultivation and the removal of harvested crops reduce the nitrogen content. Felling of trees for timber brings out a heavy outflow of nitrogen from the forest ecosystem.

An addition of commercial fertilizers increases the quantum of nitrogen in the soil. Cultivation in general increases the nitrogen content of the soil. If leguminous plants are cultivated then the increase is many times. On the whole the nitrogen cycle has to be carefully studied and any break in the ecosystem must be avoided.

Cycles of Sediments: In the sedimentary type of cycle, the major reservoir is the lithosphere from which the elements are released by weathering. The sedimentary types are best exemplified by phosphorous and sulphur. Both sulphur and phosphorous are very important factors, limiting or controlling the abundance of organisms. Both these elements have a tendency to stagnate as portion of the supply is lost as in deep sea sediments, and thereby become inaccessible to organisms and to continual cycling.

Cycle of Phosphorous: Phosphorous is an important and necessary constituent, of protoplasm as it is an essential constituent of nucleic acids, phospholipids and numerous phosphorylated compounds. The phosphorous tends to have been formed in past geological ages. These are gradually eroding, releasing phosphates to the ecosystems. Plants absorb these inorganic phosphates dissolved in water.

These phosphates are transferred to animal consumers and decomposers as organic phosphates and subsequently is made available for reçycling via mineralizing decomposition. Much phosphate is leached into the sea, where part of it is deposited in the shallow sediments and part of it is lost to the deep sediments.

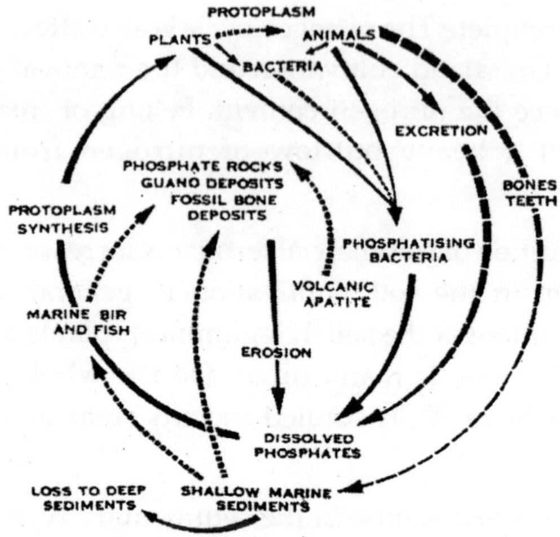

The phosphorous cycle.

Cycle of Sulphur: Sulphur is an essential constituent of certain amino acids. Not as much sulphur is required by the organisms as nitrogen and phosphorus, nor it is an often limiting to the growth of plants and animals.

Only a few organisms obtain their sulphur requirements in such organic forms as amino acids and cystine. The inorganic sulphate is the major source of biologically significant sulphur. Most of the biologically incorporated sulphur is mineralised by bacteria and fungi. Under anaerobic condition, however, hydrogen sulphide and some others are reduced direct to sulphides by bacteria as Escherichia and Proteus. Some organic sulphur enters the atmosphere as SO_2 by incomplete combustion of fossil fuels. Inorganic sulphur precipitates out as sulphate. Since sulphate is readily soluble in water, it serves as a source of elemental sulphur in many ecosystems. Sulphate is also reduced under anaerobic conditions to elemental sulphur or to sulphides by heterotrophic bacteria such as Desulfavibrio. The sulphate reducing anaerobic bacteria are heterotrophic, using the sulphate as a hydrogen acceptor in metabolic oxidation in a manner comparable to the use of nitrite and nitrate by denitrifying bacteria.

Colourless sulphur bacteria Beggiaota oxidise hydrogen sulphide to elemental sulphur and Thiobacillus oxidise it to sulphide. Certain species of Thiobacillus oxidise sulphide to sulphur and still others oxidise sulphur to sulphate. In some species oxidation process takes place in the presence of oxygen while in others presence or absence of oxygen hardly matters. These latter bacteria include the green and purple photosynthetic bacteria that use the hydrogen of H_2S as the oxygen acceptor in reducing CO_2; the green bacteria are able to oxidise sulphide to elemental sulphur, whereas the purple bacteria carry the oxidation to the sulphate stage.

The sedimentary aspect of the cycle involves the precipitation of sulphur in the presence of iron under anaerobic conditions. Ferrous sulphide is insoluble in neutral or alkaline water, and consequently sulphur has the potential for being bound up under these conditions to the limits of the iron present. Because of the thermodynamics of the ferrous sulphide system, other nutrients important to biological system can also get trapped for varying periods of time-copper, cadmium, zinc and cobalt are the examples of such nutrients. On the contrary, the very binding of these iron compounds allows the conversion of phosphorous from insoluble to soluble and thus makes it accessible. The sulphur cycle provides an excellent example of the interaction and regulation that exists between different mineral cycles.

5

Basic Resources

During early time man had to struggle for survival like animals, but now he has become the dominant species in the biosphere. He has achieved this position due to fine combination of some physical features and mental abilities. The special better character which man has developed are: (i) Large brain; (ii) Skilled hands; (iii) Power of abstract thinking; (iv) Foresight; (v) Strong memory.

Man's need are diverse e.g, food, shelter, transport, economic welfare, etc. Man is able to prepare tools and animals. He can domesticate the animals and can use their services at his will. He has devised a spoken and written language and this enables him to record and communicate his ideas and experiences to other in the form of books, etc. Mental capabilities of man have enabled him to explore and understand natural phenomena, and to manipulate the environment to his advantage. Man is more than a mere biological creature as he has a strong desire for comfortable and secure life. Therefore, he depends more on and draws more materials and energy from the environment than other organisms. He has created a socio-cultural environment which he constituted by skills, efforts, tools and other social bindings.

With the development of human civilisation, man has created new artificial ecosystems cities where soil has been covered with concrete and steel. Man also shuts himself indoor, manufacturing weather year round to suit his needs. All this finally changed the biosphere into a human dominated environment called noosphere (noo—mind, sphere—domain). Man has affected the environment form the very beginning nomad to present day civilised, i.e., hunting, fishing and food gathering stages. Man converted the self sufficient system of biosphere into a system of natural resources.

The Theory

The word resource has been defined as:

Any means of attaining given ends or

Means of supplying a material generally held in reserve.

The natural resources are the materials necessary for life and available in the normal environment such as energy, air, water, land (soil), minerals, micro organisms, plants and animals. For man, resources are the substances required for his survival and comforts. These substances are obtained directly from the environment. In fact natural resources are the components of atmosphere, hydrosphere and lithosphere.

The Classification

1. The natural resources are classified into three main types on the basis of their chemical nature:
   - (i) Inorganic resources, viz. air, water and metallic minerals.
   - (ii) Organic resources, viz. plants, animals, micro organisms and fossil fuels and
   - (iii) Mixed resources, viz. soil which is an inorganic as well as organic resource.

2. The natural resources are of three types on the basis of their distribution:
   (i) National, i.e. confined to national boundaries e.g., mineral, lands.
   (ii) Multinational, i.e. shared by many countries e.g., some rivers, certain lakes, migratory animals.
   (iii) International, i.e. used by all the nations e.g., air, solar energy.
3. With regard to their abundance and availability the natural resources are of two types:
   (i) *Inexhaustible:* These are present in plenty and are not likely to be exhausted by human use e.g., air, clay, sand, tidal energy and precipitation. Although air will never become limiting quantitatively but it may become a limiting factor qualitatively if its pollution is not checked.
   (ii) *Exhaustible:* These are likely to be finished by human use. They are further divided into:
      (a) *Renewable:* These are the resources which are continuously consumed by man but renewed by nature. They include water, soil and living objects-crops, forests, domestic animals and wild life. Some of those materials are constantly renewed in nature. Other can be reclaimed or purified and used again and again. The renewable resources are, therefore, not likely to be exhausted. However, they too may be lost by excessive and unwise use.
      (b) *Non-renewable:* They are not renewable after use and are not replenished by nature, i.e. once used are lost for ever as they are not restored e.g. metallic minerals and non-metallic mineral fuels or fossil fuels—coal, natural gas and

petroleum. All substances are regarded as renewable as matter can not be destroyed. However, coal or oil after burnt can not be used again though carbon present in air is picked up by plants. Similarly, iron of a can get rusted on a damp place in lost to us though it remains in the soil. The cycles of these substances are too long for any possible management. Therefore, such substances are said to be non-renewable. Man is making excessive use of minerals and fossil fuels is the industry and automobiles unmindful of the consequences. At current rates all the known global reserves of industrial metals may last for less than a century and those of petroleum and natural gas may exhaust in 15 to 20 years.

The Threat

Advancement in technology has led to increase in population, rise in standard of living and high expectation for better life. All these have created increased in demands on natural resources. A balance exists in nature in the relationship between organisms and their physical environment but due to increased demands and stress on natural resources man has disturbed the balance by excessive use of natural resources. This led to an ecological crisis and has threatened its existence. The only way to save man kind is the conservation of natural resources.

The Harvesting

Conservation is defined as the most efficient and most beneficial utilisation of natural resources man should take steps to ensure availability of natural resources for future generation.

(a) He should use the resources sparingly and wisely to check their depletion.

(b) He should recover the used resources for reuse as far as possible.

(c) He should discover new resources to fulfil his requirements.

(d) He should not pollute the natural resources to keep fit for use.

Energy in Power

Energy is an important input for development. It aims at human welfare covering household, agriculture transport and industrial complexes. The major energy sources at present are fuel, wood and fossil fuels. Energy resources are renewable as well as non-renewable.

(I) *Renewable (inexhaustible) Energy Resources:* These are normally biomass-based and available in unlimited amount in nature since these can be renewed over relatively short period of time. These include fire wood (or fuel wood) animal dung, solar energy, wind energy, water energy, etc.

(II) *Non-renewable (exhaustible) Energy Resources:* These are present in limited amount and take lower period to develop again. If these used unlimited, they are likely to be exhausted one day. These energy resources are coal, natural gas, mineral oil and nuclear power. Since coal, petroleum and natural gas are organic in their origin so-called fossil fuels.

Convential and Non-convential Sources: The expansion of energy sources has been directly related with the pace of industrial and agriculture development.

A. *Convential Energy Sources:* As most of the fuel wood was used for domestic purposes especially in rural areas so very little of it was available to industrial sector. Coal, already in use in industries, became a highly priced source. It was supplemented by mineral oil. Similarly, in those areas where running water and needed technology was available hydro-electriciry became dearer. After world war II nuclear power, another source of energy, was developed. All these sources are known as conventional sources of energy.

B. Non-conventional Energy Sources: As the coil and oil deposits are going to exhaust one day and most of non-renewable energy resource cause environmental pollution so we must conserve the non-renewable conventional sources and replace them by non-polluting renewable sources called non-conventional sources of energy which include urban waste, agriculture waste, energy plantations, animals and human wastes, solar energy, wind energy, tidal energy, geothermal energy, etc. these are pollution free, environmentally clean.

The requirement of energy in India are met from both commercial and non-commercial sources. Most important sources of commercial energy are coal, lignite, oil, hydroelectricity and to some extent atomic energy. Non-commercial sources of energy are firewood, agricultural wastes and animal dung. Coal is a major source of power and an industrial fuel. The estimated coal reserves of India are 123,000 million metric tonnes and four-fifth of them are confined to Jharkhand, Bihar, Bengal coal belt out of the total coal produced 75% of it is used by railways, industries and thermal power plants. In 1950-51 crude oil production was 2.5 lakh tonnes while consumption was of 31 lakh tonnes. By 1987-88 production increased to 31 million tonnes and consumption raised to 40 million tonnes.

During hunting and food gathering stages of human civilisation, the daily per capita need for energy was just 2,000-4,000 kilo calories. It was mainly derived from food and man used his own muscle power for work. During 19th century, the use of energy of fossil fuel started and the daily per capita energy increased to 70,000 kilocalories. It is interesting to note that developed countries which share 30% of total population utilise 80% of total energy.

India wastes 24% of natural gas due to inadequate facilities. Poor sections of the society use fire wood which results forming deforestation. Action of man gradually diminishing the world's forest resources. Despite depleting reserves, the demand for fossil fuels is rising rapidly. With only a few years left for depletion of petroleum stringent conservation strategies have to be adopted.

Conservation of Energy: Conservation of energy is very important because there is no life without energy. Conservation in more important than generation of energy. It is necessary to understand the importance of energy so that people get used to conserve energy. With the increase in population, energy need has also increased. In our country, the yearly demand of energy is 12.80% but the production is 9.8%. Conservation of energy means the management, proper use and saving of energy resources. The following measures can be taken to conserve energy resources.

1. Methods should be developed to reduce wastage during extraction and maximise recovery of fuel e.g. shield mining system has improved coal extraction from 35% to 70%.

2. Prospecting should be carried out in new areas to find out the existence of more mines and wells.

3. Lot of fossil fuels occur in deep mines and wells. Techniques should be developed to recover the same.

4. Wastage of fuel incurred during transportation should be avoided through use of better containers.
5. Since oil wells and coal mines are prone to catch fires efforts should be made to prevent it.
6. Fuel use efficiency of equipment, devices and vehicles should be improved with maximisation of combustion.
7. Interfuel substitution is carried out through coal gasification for producing liquid fuel and compressed natural gas (CNG). Methanol can be mixed with petrol to reduce consumption of the latter.
8. Transport system can be made to run on solar energy (e.g., solar cars). and hydrogen fuel. This will save energy and will prevent pollution.
9. Use of oil should be minimised to save it for future use.
10. Over-consumption of oil in automobiles should be checked by keeping the engines in a good working order.
11. Search for new oil reserves, on shore as well as off shore should continue more earnestly and effectively.
12. Use of biogas plants should be encouraged.
13. More fuel, wood trees and shrubs should be grown as a mass campaign.
14. Population growth should be checked to control increase in energy requirement.
15. We should useless energy and avoid wastage of energy light and fans and other home gadget should be turned on only when needed.
16. Public should be educated and about the need and modes for sewing energy.
17. More stress should be laid on alternate resources of energy like hydroelectric, geothermal, wave, tidal, solar and nuclear.

Different Sources

Energy crisis is an important environment problems of today. It is because the world's almost dependence on fossil fuel energy. There is a need to find out alternate energy resources which are renewable and non-polluting such as solar energy, hydroelectric power, wind power, geothermal nuclear, tidal, dung, garage and wood energy.

Solar Energy: It is an inexhaustible non-conventional energy resources. The amount is quite high but diffuse. Some 200 watts/m2/hr. It has a good prospects especially in tropical countries such as India. Intense efforts have been made to for its large scale harnessing. Solar equipment for space heating, water heating, air conditioning, lighting and cooking have been developed.

The solar energy resources may be converted into other form of energy through thermal or photovoltaic conversion routes. The solar thermal route uses radiations in the form of heat that in turn may be converted to mechanical, electrical or chemical energy. Solar thermal devices like solar cookers, solar water/air heaters, solar dryers have been developed. Govt gives financial help to encourage preparing solar equipment.

By photovltaic systems solar radiations can directly converted into electricity through silicon solar cells. India is going to set up its first solar power station near Delhi very soon. In Haryana, first solar water heater installed in 1982 at Haryana Breweries Ltd. Murthal. It is able to heat 15000 litres of water per day at a temp, of 65°c. Photovoltaic water pumps are also operating in India. Solar cars are a reality. The only defect in large area required for harnessing solar energy.

Wind Energy: It in an inexhaustible resource which has been used since ages for grinding grains, lifting water and propelling ships, the instruments used for harnessing wind power in called wind mill. A wind mill can be installed wherever a steady wind of more than 25km/hr is available. The force of the wind rotates the wheel which is connected to a generator or turbine for generation of electricity. Asia's largest wind mill complex in located at Lamba (Gujarat) with an installed capacity of 28 MW. The limitation of wind energy is that the wind always does not blow with required intensity all the year round and in all areas.

Hydroelectric Energy: It is virtually an inexhaustible resources. It is produced from the kinetic energy of water falling from a height/using turbines. It is the most common, non-polluting, commercial source of energy. Many hydroelectric powerhouses have established on many rivers in hilly areas of our country. In large sized hydropower plants, water in impounded in large reservoirs called dams. Dams have been used in preventing floods, regular supply of canal water and development of agriculture.

Nuclear Energy: It is a heat energy released during fusion commission of the atoms of certain elements, such as uranium. Heat energy is converted into electricity. Atomic energy (nuclear energy) is non-renewable resource. A large amount of energy is released from a very small amount of the element (1 kg of Uranium liberates an energy equivalent to 35000 kg of coal). Energy liberation is carried out in nuclear reactors. The nuclear power houses have been set up in Tarapur (Mumbai), Kota (Rajasthan) and in Narora (U.P.) Besides generating electricity, nuclear energy is also used in running submarines, warships, spacecrafts, etc.

Geothermal Energy: It is a non-conventional source of energy which is obtained from either hot rocks directly or from water trapped in hot rocks present in the interior of earth. The trapped water is changed into steam which comes out at certain places as hot springs. Steam and hot water coming out of the earth in same places can be used for heating buildings and water and for generating electricity. Already a 5 KW geothermal plant is operating at Manikaran (H.P.) However geothermal resources are limited.

Tidal Energy: It is a small but inexhaustible resource. Tidal waves of the sea can be used to generate electricity. Dual flow turbines can produce electricity both during rising and receding tide. In 1966, France constructed the first tidal electric plant. India has three good sites for tidal power plants-Gulf of Kutch, Gulf of Cambay and Sunderbans. A tidal power plant is established in Gulf of Kutch by the assistance of French.

OTEC: It is ocean thermal energy conversion. In tropical oceans there is a temperature difference of surface water (28°-30°C) and deep water (5°-7°C at a depth of 800-1000m). The difference can be used to generate electricity. Already IMW plant is operating in area of Lakshadweep Island.

Wave Energy: It is small but inexhaustible resource. A platform with sloping ramp collection basins and low head turbines is required to convert wave power into electricity. A commercial wave powered plant is operating near Bergen, Norway.

Biomass Based Energy (Renewable Resource): Biomass is the term used for all materials originating from photosynthesis. Over 3000 million tonnes of organic wastes are formed annually in India, out of them some 1650 million tonnes are animal dung, human night soil, slaughter house waste, cannery wastes, etc. The rest are crop residue, forest residue, garbage and overgrowth of quatic plants. Organic wastes are used to prepare manure, dried or directly burnt as fuel. Biogas plants yield both fuel gas and manure.

(i) *Dung Cakes:* A large quantity of animal dung in rural areas is converted into cakes and dried. The cakes are used for heating and cooking at homes in villages and small towns.

(ii) *Biogas (Gobar Gas):* Biogas is methane rich fuel gas which is produced as a product of anaerobic breakdown and fermentation of biomass. Animal dung is used to produce biogas. India has a large population of domestic animals from which about 200 million tonnes of wet dung is available. It is mixed with other organic wastes. Water content is made 90% at which most methane bacteria are active. The organic wastes are dumped in pits connected with gas storage tank. Biogas is stored in storage tank from where it supplied for cooking, heating or other purposes. The residue left forms a good manure.

Wood Energy: It is a renewable resource. Fuel wood is the most common source of energy in villages of India use of wood as a fuel is fast depleting the country's tree and forest wealth. Deforestation is causing soil erosion and floods. Energy plantation is growing of fire wood trees on non-forest, non-agricultural parts of land.

Energy Cropping: It is renewable resource which consists in growing crops from which alcohol and other energy fuels can be obtained e.g., potato, sugarcane, maize, Tapioca. Alcohol can be used as such or mixed with petrol.

Petroplants: They are renewable energy resource. These are those plants having large amount of latex with long chain liquid hydrocarbons e.g. Jatropa, Euphorbia, etc. Growing petro-plants is part of energy cropping.

Land Resources: Land is solid exposed crust of earth which occurs in 27% of the area occupying 133.93 million hectares. In India agriculture occupies nearly 43% land, forest 23%, pastures 4% and human habitations 8%. A judicious land used is important, otherwise economic disaster and ecological crisis can overtake human population any time.

Land Use Planning and Management

1. A land classification and capability map be prepared with the help of scientific survey and remote sensing method.
2. Land should be classified into types for agriculture, pastures, forests, slopes, marshes, swamps, barren, rocky, sandy, etc.
3. Projections be made for both short term and long term requirements of land for different purposes like agriculture, aquaculture, water reservoirs, forestry, grazing, human settlements, industries, road, etc.
4. Legislation should be enacted to control land use. Fertile agriculture land should not be lost to non-agricultural purposes.
5. Erosion and landslides should be prevented in hills by taking appropriate afforestation measures and soil conservation techniques.
6. Forest land should not be encroached upon.
7. Only barren and waste land should be used for building industries, residential complexes, dams and roads.
8. Solid wastes of cities and towns should not be dumped on leave fertile soil.

Soil Resources: Soil or upper fertile part of earth's crust supports life system on land. However it is prone to erosion and loss of fertility. There are several factors to cause loss of soil fertility and erosion of soil.

Reduction of Soil Fertility

(i) Over cropping withdraws most of essential minerals from upper fertile layer of the soil.

(ii) Non-rotation of crops causes efficiency of particular minerals.

(iii) Leaching of minerals.

(iv) Overgrazing.

(v) Precipitation of minerals due to change in pH.

(vi) Non-availability of organic wastes.

Restoration of Soil Fertility

(i) Stoppage of over cropping.

(ii) Rotation of crops.

(iii) Leaving the soil uncultivated for 1-2 years.

(iv) Addition of manure and compost.

(v) Green maneuring.

(vi) Addition of suitable fertilizers.

(vii) Avoiding irrigation soils with brackish water.

(viii) Reduction of soil silnity through leaching of salts.

Water Resources: Earth is a water planet with 77% of its surface covered with water. Water is an inorganic and renewable resource which is an essential component of all living beings and habitat of several organisms. Water circulates between earth and its atmosphere. Water is of three types fresh brackish and marine.

Fresh Water Resources: Fresh water constitutes only 2.7% of total water content of the earth. Out of this 77.2% is stored in glaciers and ice caps. And 22.4% is ground water and solid moisture. Remaining 0.36% is found in lakes, rivers, streams and swamps, etc.

The ultimate source of freshwater in rainfall India revives about 2750 km^3 of rainfall every year. About 600 km^3 seeps into ground while 900 km^3 water evaporates. Ground water is major source of fresh water. It is neither used by plants nor exposed to evaporation. However it can be pumped out.

The Uses

1. Fresh water is used for drinking by human beings. It is also used in cooking, washing, bathing, disposal of sewage.
2. Maximum water is consumed in agriculture.
3. Industries use a large amount of water for cooling, heating and other purpose.
4. Impounding of water and generation of electricity.
5. Water ways are used for inland transport.
6. Construction of buildings.
7. Rearing of fish and other aquatic organisms called aquaculture.

Fresh water resources facing a number of problems due to.

1. *Unequal Distribution:* Though 113 trillion m^3 of rain water falls annually on earth but water availability is not same everywhere.
2. *Increasing Water Use:* Since 1950, per capita use of fresh water has trebled in India.
3. *Growing Cities:* As all cities are growing very fast the demand of water is also increasing.
4. *Depleting Ground Water:* Due to excessive use of water table has lowered at number of places.
5. *Agriculture:* In India irrigation is assured in only 45% of the cropland. The remainder has to be provided with irrigation water.
6. *Waste Water:* A lot of waste water is produced by industries and domestic users which is passed into water bodies causing pollution.

Management and Conservation

(i) *Treatment of Waste Water:* The waste water must be rid

of its inorganic and organic ingredients through primary and secondary treatments.

(ii) *Use of Treated Waste Water:* The treated water should not be passed into rivers, lakes but should be used for irrigation.

(iii) *Drinking Water:* For a good management of fresh water resource, there should be proper arrangement for supply of drinking water.

(iv) *Afforestation and Reforestation:* Water availability-increases if hilis, watersheds, catechumen areas and slops are covered by vegetation.

(v) *Impoundments:* Dams and reservoirs built upstream or downstream can help reduce intensity of floods, make water available to rivers and canals.

(vi) Reducing over consumption.

(vii) *Linking River Systems:* This can help send water from a region of surplus to the area of scarcity.

Marine or Ocean Resources: Oceans are the largest reservoirs of water having over 97% of total free water and occupying 70% of earth's surface. Oceans produce more biomass than the land, function as heat bank and help maintain atmospheric balance of carbon dioxide by absorbing the extra one.

As the food and energy available on land is continuously decreasing, scientists attention has been diverted towards oceans which can be used in the following ways:

(i) As source of food and other edible animals mainly fishes.

(ii) As a source of edible sea weeds such as many brown algal (kelps), etc.

(iii) For the production of useful products like algin, agar, carrageenan, diatomitic, etc.

(iv) For generating power from tides.

(v) For extraction of common salts.

(vi) Planning floating cities on ocean surface is under way.

(vii) To build shore terminals for ships, etc.

(viii) It is used for commercial navigation and shipping.

(ix) Ocean floor contains polymetallic nodules commonly called managese nodules.

Problems

(i) *Silt:* Over 3000 million tonnes of silt enters the sea from land through rivers.

(ii) *Wastes:* Towns, cities and industries situated on the sea coast pass their effluents, sewage and garbage in sea.

(iii) *Toxic Wastes:* There is a tendency to dump toxic wastes in sealed containers deep in sea which soon leak and release the waste.

(iv) *Eutrophication:* Due to pouring of wastes in sea, electrification, (bloom forming algae) occurs which are toxic and kill the animals.

Management and Conservation

(i) *Treatment of Wastes:* Waste and waste water should neither be dumped directly in oceans nor in rivers. They must be treated.

(ii) *Oil Slicks and Spills:* They should be immediately cleaned.

(iii) *Fish Catch:* It should be regulated to prevent reduction in fish population.

(iv) *Mariculture:* It should be undertaken on large scale but only for providing food but also marine products like pearls.

Mineral Resources: They are inorganic substances obtained or mined from earth's crust. Minerals were deposited only once during the cooling of earth's crust. Fresh deposits cannot occur except through volcanic eruption. An orebody is an area in the crust where the minerals occur in such a concentration that it can be economically exploited. The amount of minerals found in an ore is called resource. The various steps involved in minerals utilisation are:

(i) The mineral extraction.

(ii) Conversion of minerals into bulk metal, ceramics and chemicals.

(iii) Manufacture of public utility goods such as medicines, metallic structures, household utensils and pesticides.

(iv) Their return to environment as waste refuse.

Wealth under Earth

(i) *Mining:* Wastage during mining should be minimised.

(ii) *Poor Mines:* Techniques should be developed to extract mineral from mines of low quality.

(iii) *Transport:* Wastage during ore transport should be prevented.

(iv) *Purification:* It involves a lot of wastage which should be reduced to minimum.

(v) *Effluents:* Areas of all minerals contain smaller quantities of other minerals. They should not be allowed to go waste but reclaimed.

(vi) *Recycling:* It is recirculating a material recycling reduces pressure on mining. It is also cheaper because of lesser energy requirement.

(vii) *Substitution:* It is replacement of a mineral with another in some of the uses.

6

Energy Sources

Energy consumption of a nation is usually considered as an index of its development. This is because almost all the developmental activities are directly or indirectly dependent upon energy. We find wide disparities in per capita energy use between the developed and the developing nations.

The first form of energy technology probably was the fire, which produced heat and the early man used it for cooking and heating purposes. Wind and hydropower have also been in use for the last 10,000 years. The invention of steam engines replaced the burning of wood by coal and coal was later replaced to a great extent by oil. In 1970's due to Iranian revolution and Arab oil embargo the prices of oil shooted up. This ultimately led to exploration and use of several alternate sources of energy.

Energy: Urgent Requirements: Development in different sectors relies largely upon energy. Agriculture, industry, mining, transportation, lighting, cooling and heating in buildings all need energy. With the demands of growing population the world is facing further energy deficit. The fossil fuels like coal, oil and natural gas which at present are supplying 95% of the commercial energy of the world resources and are not going to last for many more years. Our life-style is changing very fast and from a simple way of life we are shifting to a luxurious life-style. If you just look at the number of electric gadgets in your homes and the number of private cars and scooters in your locality you will realise that in the last few years they have multiplied many folds and all of them consume energy.

Developed countries like U.S.A. and Canada constitute about 5% of the world's population but consume one fourth of global energy resources. An average person there consumes 300 GJ (Giga Joules, equal to 60 barrels of oils) per year. By contrast, an average man in a poor country like Bhutan, Nepal or Ethiopia consumes less than 1 GJ in a year. So a person in a rich country consumes almost as much energy in a single day as one person does in a whole year in a poor country. This clearly shows that our life-style and standard of living are closely related to energy needs. U.S.A., Norway, Switzerland, etc. with high GNP show high energy use while India, China, etc. have low GNP and low energy use. Behrain and Qatar are oil rich states (UAE) and hence their energy consumption and GNP are more, although their development is not that high.

Different Energy Sources

A source of energy is one that can provide adequate amount of energy in a usable form over a long period of time. These sources can be of two types:

(1) *Renewable Resources:* which can be generated continuously in nature and are inexhaustible e.g. wood, solar energy, wind energy, tidal energy, hydropower, Bio-mass energy, bio-fuels, geothermal energy and hydrogen. They are also known as non-conventional sources of energy and they can be used again and again in an endless manner.

(2) *Non-renewable Resources:* which have accumulated in nature over a long span of time and cannot be quickly

replenished when exhausted e.g. coal, petroleum, natural gas and nuclear fuels like uranium and thorium.

Wood is a renewable resource as we can get new wood by growing a sapling into a tree within 15-20 years but it has taken millions of years for the formation of coal from trees and cannot be regenerated in our life time, hence coal is not renewable. We will now discuss various forms of renewable and non-renewable energy resource.

Renewable Energy Resources

Solar Energy: Sun is the ultimate source of energy, directly or indirectly for all other forms of energy. The nuclear fusion reactions occurring inside the sun release enormous quantities of energy in the form of heat and light. The solar energy received by the near earth space is approximately 1.4 kilojoules/second/m^2 known as solar constant.

Traditionally, we have been using solar energy for drying clothes and food-grains, preservation of eatables and for obtaining salt from sea-water. Now we have several techniques for harnessing solar energy. Some important solar energy harvesting devices are discussed here.

(i) *Solar Heat Collectors:* These can be passive or active in nature. Passive solar heat collectors are natural materials like stones, bricks, etc. or material like glass which absorb heat during the day time and release it slowly at night. Active solar collectors pump a heat absorbing medium (air or water) through a small collector which is normally placed on the top of the building.

(ii) *Solar Cells:* They are also known as photovoltaic cells or PV cells. Solar cells are made of thin wafers of semi conductor materials like silicon and gallium. When solar radiations fall on them, a potential difference is produced which causes flow of electrons and produces electricity. Silicon can be obtained from silica or sand, which is abundantly available and inexpensive. By using gallium arsenide, cadmium sulphide or boron, efficiency of the PV cells can be improved. The potential difference produced by a single PV cell of 4 cm^2 size is about 0.4-0.5 volts and produces a current of 60 milli amperes.

A group of solar cells joined together in a definite pattern form a solar panel which can harness a large amount of solar energy and can produce electricity enough to run street-light, irrigation water pump, etc.

Solar cells are widely used in calculators, electronic watches, street lighting, traffic signals, water pumps, etc. They are also used in artificial satellites for electricity generation. Solar cells are used for running radio and television also. They are more in use in remote areas where conventional electricity supply is a problem.

(iii) *Solar Cooker:* Solar cookers make use of solar heat by reflecting the solar radiations using a mirror directly on to a glass sheet which covers the black insulated box within which the raw food is kept. A new design of solar cooker is now available which involves a spherical reflector (concave or parabolic reflector) instead of plane mirror that has more heating effect and hence greater efficiency.

The food cooked in solar cookers is more nutritious due to slow heating. However it has the limitation that it cannot be used at night or on cloudy days. Moreover, the direction of the cooker has to be adjusted according to the direction of the sun rays.

(iv) *Solar Water Heater:* It consists of an insulated box painted black from inside and having a glass lid to receive and store solar heat. Inside the box it has black painted copper coil through which cold water is made to flow in, which gets heated and flows out into a storage tank.

The hot water from the storage tank fitted on roof top is then supplied through pipes into buildings like hotels and hospitals.

(v) *Solar Furnace:* Here thousands of small plane mirrors are arranged in concave reflectors, all of which collect the solar heat and produce as high a temperature as 3000°C.

(vi) *Solar Power Plant:* Solar energy is harnessed on a large scale by using concave reflectors which cause boiling of water to produce steam. The steam turbine drives a generator to produce electricity. A solar power plant (50 K Watt capacity) has been installed at Gurgaon, Haryana.

Wind for Energy: The high speed winds have a lot of energy in them as kinetic energy due to their motion. The driving force of the winds is the sun. The wind energy is harnessed by making use of wind mills. The blades of the wind mill keep on rotating continuously due to the force of the striking wind.

The rotational motion of the blades drives a number of machines like water pumps, flour mills and electric generators. A large number of wind mills are installed in clusters called wind farms, which feed power to the utility grid and produce a large amount of electricity. These farms are ideally located in coastal regions, open grasslands or hilly regions, particularly mountain passes and ridges where the winds are strong and steady. The minimum wind speed required for satisfactory working of a wind generator is 15 km/hr. The wind power potential of our country is estimated to be about 20,000 MW, while at present we are generating about 1020 MW. The largest wind farm of our country is near Kanyakumari in Tamil Nadu generating 380 MW electricity.

Wind energy is very useful as it does not cause any air pollution. After the initial installation cost, the wind energy is very cheap. It is believed that by the middle of the century wind power would supply more than 10% of world's electricity.

Hydropower: The water flowing in a river is collected by constructing a big dam where the water is stored and allowed to fall from a height. The blades of the turbine located at the bottom of the dam move with the fast moving water which in turn rotate the generator and produces electricity. We can also construct mini or micro hydel power plants on the rivers in hilly regions for harnessing the hydro energy on a small scale, but the minimum height of the water falls should be 10 metres. The hydropower potential of India is estimated to be about 4×10^{11} KW-hours. Till now we have utilised only a little more than 11% of this potential.

Hydropower does not cause any pollution, it is renewable and normally the hydro power projects are multi-purpose projects helping in controlling floods, used for irrigation, navigation, etc. However, big dams are often associated with a number of environmental impacts which have already been discussed in the previous section.

Tidal Energy: Ocean tides produced by gravitational forces of sun and moon contain enormous amounts of energy. The 'high tide' and 'low tide' refer to the rise and fall of water in the oceans. A difference of several meters is required between the height of high and low tide to spin the turbines. The tidal energy can be harnessed by constructing a tidal barrage. During high tide, the sea-water flows into the reservoir of the barrage and turns the turbine, which in turn, produces electricity by rotating the generators. During low tide, when the sea-level is low, the sea water stored in the barrage reservoir flows out into the sea and again turns the turbines.

There are only a few sites in the world where tidal energy can be suitably harnessed. The bay of Fundy Canada having 17-18 m high tides has a potential of 5,000 MW of power generation. The tidal mill at La Ranee, France is one of the first modern tidal power mill. In India Gulf of Cambay, Gulf of Kutch and the Sunder bans deltas are the tidal power sites.

Ocean Thermal Energy (OTE): The energy available due to the difference in temperature of water at the surface of the tropical oceans and at deeper levels is called Ocean Thermal Energy. A difference of 20°C or more is required between surface water and deeper water of ocean for operating OTEC (Ocean Thermal Energy Conversion) power plants. The warm surface water of ocean is used to boil a liquid like ammonia. The high pressure vapours of the liquid formed by boiling are then used to turn the turbine of a generator and produce electricity. The colder water from the deeper oceans is pumped to cool and condense the vapours into liquid. Thus the process keeps on going continuously for 24 hours a day.

Geothermal Energy: The energy harnessed from the hot rocks present inside the earth is called geothermal energy. High temperature, high pressure steam fields exist below the earth's surface in many places. This heat comes from the fission of radioactive material naturally present in the rocks. In some places, the steam or the hot water comes out of the ground naturally through cracks in the form of natural geysers as in Manikaran, Kullu and Sohana, Haryana. Sometimes the steam or boiling water underneath the earth do not find any place to come out. We can artificially drill a hole up to the hot rocks and by putting a pipe in it make the steam or hot water gush out through the pipe at high pressure which turns the turbine of a generator to produce electricity. In USA and New Zealand, there are several geothermal plants working successfully.

Greatest Source

Bio-mass is the organic matter produced by the plants or animals which include wood, crop residues, cattle dung, manure, sewage, agricultural wastes, etc. Bio-mass energy is of the following types:

(a) *Energy Plantations:* Solar energy is trapped by green plants through photosynthesis and converted into Bio-mass energy. Fast growing trees like cottonwood, poplar and *Leucaena*, non-woody herbaceous grasses, crop plants like sugarcane, sweet sorghum and sugar beet, aquatic weeds like water hyacinth and sea-weeds and carbohydrate rich potato, cereal, etc. are some of the important energy plantations. They may produce energy either by burning directly or by getting converted into burnable gas or may be converted into fuels by fermentation.

(b) *Petro-crops:* Certain latex-containing plants like *Euphorbias* and oil palms are rich in hydrocarbons and can yield an oil like substance under high temperature and pressure. This oily material may be burned in diesel engines directly or may be refined to form gasoline. These plants are popularly known as petro-crops.

(c) *Agricultural and Urban Waste Bio-mass:* Crop residues, bagasse (sugarcane residues), coconut shells, peanut hulls, cotton stalks, etc. are some of the common agricultural wastes which produce energy by burning. Animal dung, fishery and poultry waste and even human refuse are examples of Bio-mass energy. In Brazil 30 % of electricity is obtained from burning bagasse. In rural India, animal dung cakes are burnt to produce heat. About 80 % of rural heat energy requirements are met by burning agricultural wastes, wood and animal dung cakes.

In rural areas these forms of waste Bio-mass are burned in open furnaces called 'Chulhas' which usually produce smoke and are not so efficient (efficiency is <8 %). Now improved Chulhas with tall chimney have been designed which have high efficiency and are smokeless. The burning of plant residues or animal wastes cause air pollution and produce a lot of ash as waste residue. The burning of dung destroys essential nutrients like N and P. It is therefore, more useful to convert the Bio-mass into Bio-gas or bio-fuels.

Bio-gas: Bio-gas is a mixture of methane, carbon dioxide, hydrogen and hydrogen sulphide, the major constituent being methane. Bio-gas is produced by anaerobic degradation of animal wastes (sometimes plant wastes) in the presence of water. Anaerobic degradation means breakdown of organic matter by bacteria in the absence of oxygen.

Bio-gas is a non-polluting, clean and low-cost fuel which is very useful for rural areas where a lot of animal waste and agricultural waste are available. India has the largest cattle population in the world (240 million) and has tremendous potential for Bio-gas production. From cattle dung alone, we can produce Bio-gas of a magnitude of 22,500 Mm^3 annually. A sixty cubic feet gobar gas plant can serve the needs of one average family.

Bio-gas has the following main advantages: It is clean, non-polluting and cheap. There is direct supply of gas from the plant and there is no storage problem. The sludge left over is a rich fertilizer containing bacterial bio-mass with most of the nutrients preserved as such. Air-tight digestion/degradation of the animal wastes is safe as it eliminates health hazards which normally occur in case of direct use of dung due to direct exposure to faecal pathogens and parasites.

Bio-gas plants used in our country are basically of two types: 1. Floating gas-holder type and 2. Fixed-dome type.

1. *Floating Gas Holder type Bio-gas Plant:* This type has a well-shaped digester tank which is placed under the ground and made up of bricks. In the digester tank, over the dung slurry an inverted steel drum floats to hold the bio-gas produced. The gas holder can move which is controlled by a pipe and the gas outlet is

regulated by a valve. The digester tank has a partition wall and one side of it receives the dung-water mixture through inlet pipe while the other side discharges the spent slurry through outlet pipe.

Sometimes corrosion of steel gas-holder leads to leakage of Bio-gas. The tank has to be painted time and again for maintenance which increases the cost. Hence another type was designed as discussed below:

2. *Fixed Dome type Bio-gas Plant:* The structure is almost similar to that of the previous type. However, instead of a steel gas-holder there is dome shaped roof made of cement and bricks. Instead of partitioning, here there is a single unit in the main digester but it has inlet and outlet chamber.

The Ministry of Non-Conventional Energy Sources (MNES) has been promoting the Bio-gas Programme in India. Out of the various models, the important ones used in rural setup are KVIC Model (Floating drum type), Janta Model (Fixed dome type), Deenbandhu Model (Fixed dome type), Pragati Model (floating drum type), Ganesh Model (KVIC type but made of bamboo and polythene sheet) and Ferro-cement digester Model (KVIC type with ferro-cement digester).

Bio-fuels: Bio-mass can be fermented to alcohols like ethanol and methanol which can be used as fuels. Ethanol can be easily produced from carbohydrate-rich substances like sugarcane. It burns clean and is non-polluting. However, as compared to petrol its calorific value is less and therefore, produces much less heat than petrol.

Gasohol is a common fuel used in Brazil and Zimbabwe for running cars and buses. In India too gasohol is planned to be used on trial basis in some parts of the country, to start with in Kanpur. Gasohol is a mixture of ethanol and gasoline. Methanol is very useful since it burns at a lower temperature than gasoline or diesel. Thus the bulky radiator may be substituted by sleek designs in our cars. Methanol too is a clean, non-polluting fuel.

Methanol can be easily obtained from woody plants and ethanol from grain-based or sugar-containing plants.

Hydrogen as a Fuel: As hydrogen burns in air, it combines with oxygen to form water and a large amount of energy (150 kilojoules per gram) is released. Due to its high, rather the highest calorific value, hydrogen can serve as an excellent fuel. Moreover, it is non-polluting and can be easily produced. Production of Hydrogen is possible by thermal dissociation, photolysis or electrolysis of water:

(i) By thermal dissociation of water (at 3000°K or above) hydrogen (H_2) is produced.

(ii) Thermochemically, hydrogen is produced by chemical reaction of water with some other chemicals in 2-3 cycles so that we do not need the high temperatures as in direct thermal method and ultimately H_2 is produced.

(iii) Electrolytic method dissociates water into hydrogen (H_2) and oxygen by making a current flow through it.

(iv) Photolysis of water involves breakdown of water in the presence of sun light to release hydrogen. Green plants also have photolysis of water during photosynthesis. Efforts are underway to trap hydrogen molecule which is produced during photosynthesis.

However, hydrogen is highly inflammable and explosive in nature. Hence, safe handling is required for using H_2 as a fuel. Also, it is difficult to store and transport. And, being very light, it would have to be stored in bulk. Presently, H_2 is used in the form of liquid hydrogen as a fuel in spaceships.

Non-Renewable Energy Sources: These are the fossil fuels like coal, petroleum, natural gas and nuclear fuels. These were formed by the decomposition of the remains of plants and animals buried under the earth millions of years ago. The fuels are very precious because they have taken such a long time to be formed and if we exhaust their reserves at such a fast rate as we have been doing ever since we discovered them, then very soon we will lose these resources forever.

Coal: Coal was formed 255-350 million years ago in the hot, damp regions of the earth during the carboniferous age. The ancient plants along the banks of rivers and swamps were buried after death into the soil and due to the heat and pressure gradually got converted into peat and coal over millions of years of time. There are mainly three types of coal, namely *anthracite (hard* coal), *bituminous* (Soft coal) and *lignite* (brown coal). Anthracite coal has maximum carbon (90%) and calorific value (8700 kcal/kg.) Bituminous, lignite and peat contain 80, 70 and 60% carbon, respectively. Coal is the most abundant fossil fuel in the world. At the present rate of usage, the coal reserves are likely to last for about 200 years and if its use increases by 2% per year, then it will last for another 65 years.

India has about 5% of world's coal and Indian coal is not very good in terms of heat capacity. Major coal fields in India are Raniganj, Jharia, Bokaro, Singrauli, and Godavari valley. The coal states of India are Jharkhand, Orissa, West Bengal, Madhya Pradesh, Andhra Pradesh and Maharashtra. Anthracite coal occurs only in J & K.

When coal is burnt it produces carbon dioxide, which is a greenhouse gas responsible for causing enhanced global warming. Coal also contains impurities like sulphur and therefore as it burns the smoke contains toxic gases like oxides of sulphur and nitrogen.

Petroleum: It is the lifeline of global economy. There are 13 countries in the world having 67% of the petroleum reserves which together form the OPEC (Organisation of Petroleum

Exporting Countries). About 1/4th of the oil reserves are in Saudi Arabia.

At the present rate of usage, the world's crude oil reserves are estimated to get exhausted in just 40 years. Some optimists, however, believe that there are some yet undiscovered reserves. Even then the crude oil reserves will last for another 40 years or so. Crude petroleum is a complex mixture of alkane hydrocarbons. Hence it has to be purified and refined by the process of fractional distillation, during which process different constituents separate out at different temperatures. We get a large variety of products from this, namely, petroleum gas, kerosene, petrol, diesel, fuel oil, lubricating oil, paraffin wax, asphalt, plastic, etc.

Petroleum is a cleaner fuel as compared to coal as it burns completely and leaves no residue. It is also easier to transport and use. That is the reason why petroleum is preferred amongst all the fossil fuels. Liquefied petroleum gas (LPG): The main component of petroleum is butane, the other being propane and ethane. The petroleum gas is easily converted to liquid form under pressure as LPG. It is odourless, but the LPG in our domestic gas cylinders gives a foul smell. This is, in fact, due to ethyl mercaptan, a foul-smelling gas, added to LPG so that any leakage of LPG from the cylinder can be detected instantaneously. Oil fields in India are located at Digboi (Assam), Gujarat Plains and Bombay High, offshore areas in deltaic coasts of Godavari, Krishna, Kaveri and Mahanadi.

Natural Gas: It is mainly composed of methane (95%) with small amounts of propane and ethane. It is a fossil fuel. Natural gas deposits mostly accompany oil deposits because it has been formed by decomposing remains of dead animals and plants buried under the earth. Natural gas is the cleanest fossil fuel. It can be easily transported through pipelines. It has a high calorific value of about 50KJ/G and burns without any smoke.

Currently, the amount of natural gas deposits in the world are of the order of 80, 450 gm^{-3}. Russia has maximum reserves (40%), followed by Iran (14%) and USA (7%). Natural gas reserves are found in association with all the oil fields in India. Some new gas fields have been found in Tripura, Jaisalmer, off-shore area of Mumbai and the Krishna Godavari Delta.

Natural gas is used as a domestic and industrial fuel. It is used as a fuel in thermal power plants for generating electricity. It is used as a source of hydrogen gas in fertilizer industry and as a source of carbon in tyre industry.

Compressed Natural Gas (CNG): It is being used as an alternative to petrol and diesel for transport of vehicles. Delhi has totally switched over to CNG where buses, auto-rickshaws and taxi-cars run on this new fuel. CNG use has greatly reduced vehicular pollution in the city.

Synthetic Natural Gas (SNG): It is a mixture of carbon monoxide and hydrogen. It is a connecting link between a fossil fuel and substituted natural gas. Low-grade coal is initially transformed into synthetic gas by gasification followed by catalytic conversion to methane.

Atomic Sides

Nuclear energy is known for its high destructive power as evidenced from nuclear weapons. The nuclear energy can also be harnessed for providing commercial energy. Nuclear energy can be generated by two types of reactions:

(i) *Nuclear Fission:* It is the nuclear change in which nucleus of certain isotopes with large mass numbers are split into lighter nuclei on bombardment by neutrons and a large amount of energy is released through a chain reaction.

$${}_{92}U^{235} + {}_{0}n^{1} \text{ C } {}_{36}Kr^{92} + {}_{56}Ba^{141} + 3\ {}_{0}n^{1} + \text{Energy}$$

Nuclear Reactors make use of nuclear chain reaction. In order to control the rate of fission, only 1 neutron released is allowed to strike for splitting another nucleus. Uranium-235 nuclei are most commonly used in nuclear reactors.

(ii) *Nuclear Fusion:* Here two isotopes of a light element are forced together at extremely high temperatures (1 billion °C) until they fuse to form a heavier nucleus releasing enormous energy in the process. It is difficult to initiate the process but it releases more energy than nuclear fission.

$${}_1H^2 + {}_1H^2 \rightarrow {}_3He^2 + {}_0n^1 + \text{Energy}$$

Two hydrogen-2 (Deuterium) atoms may fuse to form the nucleus of Helium at 1 billion °C and release a huge amount of energy. Nuclear fusion reaction can also take place between one Hydrogen-2 (Deuterium) and one Hydrogen-3 (Tritium) nucleus at 100 million °C forming Helium-4 nucleus, one neutron and a huge amount of energy.

Nuclear energy has tremendous potential but any leakage from the reactor may cause devastating nuclear pollution. Disposal of the nuclear waste is also a big problem.

Nuclear power in India is still not very well developed. There are four nuclear power stations with an installed capacity of 2005 MW.

These are located at Tarapur (Maharashtra), Rana Pratap Sagar near Kota (Rajasthan), Kalpakkam (Tamil Nadu) and Narora (U.P.).

7

Nature at Work

Iron

- Approximately 65% of the total body iron is present in the form of haemoglobin.
- Widely iron is utilised in the body for various functions such as
  - (i) For transport of oxygen to the issue.
  - (ii) For maintenance of oxidative enzyme system within the tissue cells.
  - (iii) It is also concerned in melanin formation.

Sources of Iron: Various sources of iron like—Green leafy materials, most leguminous plants and seed coats are well sources.

Deficiency Symptoms of Iron

(i) By the deficiency of iron decrease growth rate.

(ii) Skin colour may be redden

(iii) The deficiency of iron is effect the formation of haemoglobin compound.

Zinc

- Zinc has been found in every tissue in the animal body.
- Sources of Zinc: The chief sources is wheat meal, safflower seed oil meal, fish meal, etc.
- Deficiency Symptoms of Zinc: (i) Growth retarded; (ii) Disorders of the feathers and hair coat; (iii) Skin disease increase with the deficiency of zinc; (iv) Disorders of bones increase; (v) Reduced efficiency of feed Utilisation.

Copper: Even copper is very essential for the body and various functions in the body like for haemoglobin formation.

Deficiency of Copper

(i) Bone defects in grazing cattle and sheep on copper deficient.

(ii) Severe diarrhoea has been observed in many parts of the world by the deficiency of copper.

(iii) In sheep, copper deficiency produces a lack of black wooled sheep and characteristic less of "Crimp" form the fibers of all wool.

Manganese

- Even manganese are essential element for various functions in body.
- the highest concentration of manganese is in bone followed by pituitary gland, pineal gland, lactating mammary gland, liver gastrointestinal tissue, kidney and pancreas.

Deficiency Symptoms of Manganese

(i) In chickens the most dramatic mangnese deficiency syndrome occurs which is termed as "perosis" and is characterised by gross enlargement and malformation of the tibiometatarsal joint.

(ii) By the deficiency of Mn, reduced growth, Slightly reduced mineralisation, defective structure of bone.

Theory of Vitamins

- A vitamin is a organic compound which is present in normal food in minute amounts.
- Vitamins is essential for development of normal tissue and for normal health, growth and maintenance.
- When vitamins absent in the diet or not properly utilised, causes a specific deficiency disease or syndrome.
- Vitamins can not be synthesised by the host and therefore must be obtained either from the diet or from the micro-organisms of the intestinal tract.
- The vitamins are generally divided into two major groups: fat soluble which is usually found associated with the lipids of natural foods include vitamins A, D, E and K and water Soluble

Fat Soluble Vitamins

- The members of this group are soluble in fat Solvents such as chloroform diethyl ether, petroleum ether, etc. They are found in nature closely associated with those tissues that tend to store fats and oils.

Vitamin A

- Vitamin A was identified by Stechback.
- Vitamin A is dietary essential for all animals. Vitamin a does not occur in plants as such but it occurs as it precursor carotene. Carotene is known as provitamin A as it is converted into vitamin A inside the body.
- Vitamin A also known as retinol or retinal or retinoic acid.

Deficiency symptoms of Vitamin: various symptoms occurred in cattle and pigs like skin conditions, xerophthalmia through the deficiency of vitamin A. Poultry also affected by deficiency of vitamin A such as retarded growth, high mortality.

- Vitamin A is also responsible for the normal development of bones as it controls the activity of the osteoblast and osteoclasts of the epithelial cartilage.

Vitamin D

- Even vitamin D fat soluble.
- Vitamin D is also known as antirachitic factor as it prevents rachitis.
- Rickets is caused due to lack of normal calcification so that vitamin D is necessary for normal calcification.
- The disease rachitis, now commonly called rickets.
- Vitamin D is also involved to facilitate the clearance of phosphate in the kidney.

Deficiency Symptoms

(i) In older animals vitamin D deficiency causes ostemalacia, where there is reabsorptionof bone already laid down.

(ii) In young animals vitamin D deficiency results in rickets and retarded growth.

(iii) In Poultry, a deficiency of vitamin D causes the bones and beak to become soft and rubbery; growth is usually retarded and the legs may become bowed.

Sources of Vitamin D: Vitamin D is animal kingdom it is abundant in fish liver on entire body oil. Fresh green alfalfa forage has zero vitamin D value where as sun cured hay contain on an average 200 I.U. per 100 gms.

Vitamin E

- Mattil and Conklin (1920) indicated that rats fed on a milk diet supplemented with yeast and iron were unable to bear young.
- In 1922 Bishop and Evans announced the existance of a factor X in certain foods for normal rat reproduction.
- In the year 1936, Evans and his colleagues isolated pure vitamin E from the unsaponifiable fraction of wheat-germ oil.
- The active substance was later termed by Evans as "Vitamin E" and now at least eight compounds with E activity are known to occur in a variety of plant and animal tissues. These are called "tocopherols".
- Vitamin E acts as a antioxidant in the cellular level. Thus for example it prevents the oxidation of unsaturated fatty acids, mostly present in all cell wall components.
- Vitamin E also participates in normal tissue respiration.
- Vitamin E is also involved in the synthesis of ascorbic acid and co-enzyme and in the metabolism of uncleic acid (DNA) probably by regulating the incorportaion of pyrimidines into the nucleic acid structure and sulphur amino acid.

The Deficiency Symptoms of Vitamin E: Most animals fail to reproduce and young cattle and lambs muscle degeneration.

Sources of Vitamin E: Viamin E is widely distributed in live stock feed. It is found abundantly in whole cereal grains mainly in the germ and therefore in the by-products containing the germ.

- Leafy materials, green forage, good quality hay are very good sources.
- Alfalfa is very good sources of vitamin E.

Vitamin K

- Vitamin K was identified in 1935 by Henirk Dam as a factor present in green leaves which prevented a hemorrhagic syndrome observed in chicks maintained on a cow fat diet.
- Vitamin K is an essential metabolite for humans and for farm animals.
- Vitamin K helps in the formation of prothrombin which is necessary for normal blood coagulation.
- The deficiency of this factor lowers the prothrombin content of the blood.
- Prothrombin formation takes place in liver which is assisted by vitamin K.
- The prolongation of clotting time of the blood is due to lack of vitamin K.

Sources of Vitamin K: All green leaf plant materials, dry or fresh are rich sources of the vitamin.

Water Soluble Vitamins

- Ascorbic acid was first isolated by Szent Gyorgyi (1928) from orange juice, cabbage juice and adrenal cortex; he named the compound hexuronic acid in recognition of the six carbon atom in the molecule
- Vitamin C is chemically known as L-ascorbic acid.
- Vitamin C highly soluble in water and slightly soluble in alcohol.
- Vitamin C is stored in all tissues with the highest concentrations in the adrenal and pituitary glands and lower levels in the liver, spleen and brain.
- Vitamin C is essential for the collagen formation.
- Vitamin C aids for the conversion of folic acid to its active form tetrahydrofolic acid.

- Vitamin C is also involved in the hydroxylation of proline, lysine and aniline which are important for normal physiology of the animal.
- It involves in the metabolism of lipids as blood cholesterol level appear to fall with the administration of ascorbic acid and rise due to deficiency of vitamin.
- It aid for the conversion of tryptophan to serotonin.

Deficiency Symptoms: Deficiency symptoms for form animals is unknown for man, monkey and guinea pigs extensive study has been made.

- A deficiency of this vitamin has been found to produce a disease known as scurvy which is characterised by Swollen, bleeding and ulcerated gums, loosening of teeth, weak bones and fragility of capillaries resulting in haemorrhages throughout the body.

Sources: Citrus fruits, tomatoes, green leafy vegetables, potatoes are certain other fruits and vegetables,

Vitamin B_1 (Thiamine)

- In 1926, Jansen and Donath, successor of Eijkmann succeeded in crystalizing vitamin B_1 in pure form. The identification of the structure of the vitamin and its synthesis was accomplished in 1936 by Williams and co-workers in the same laboratory.

Deficiency Symptoms: The deficiency of thiamine in humans occurs first as a numbness of the legs, later with pain in the calf muscle, severe exhaustion, finally emaciation and paralysis.

- The patient has difficulty in breathing, there is an abnormal enlargement of the right side of the heart and a decreases in the rate of the heart beat.
- Symptoms in animals termed polyneuritis are particularly characterised by unthrifitiness are paralysis convulsions and in birds like pigeons, chicks, etc.

Paralysis appears to be due to accumulation in the brain and muscles of lactic acid.

Sources: Thiamine has widespread distribution in food but some food are very richest sources of this vitamin. In cereal grains most of the thiamine is in the outer grains layers. Green leafy vegetables bean are good sources. Animal products rich in thiamine include egg yolk liver, kidney, etc.

Vitamin B_2 (Riboflavin)

- This is the IInd member of the vitamin B complex. This is the sparingly soluble in water but differ from thiamine in being comparatively stable to heat.

Riboflavin was synthesised independently by Khun etal (1935) at Heidelberg.

Functions: Riboflavin in the form of flavin monoumcleotides (FMN) and flavin adinine dinucleotide (FAD) acts as the prosthetic group of several enzymes involved in biological oxidation reduction reactions.

Deficiency Symptoms: The most characteristic symptom of riboflavin deficiency in chicks is leg paralysis or curled toe paralysis and other common symptoms are diarrhoea, low egg production and poor hatchibility in laying birds.

- In swine the symptoms are stiffched limbs, thickned skin, skin eruptions, lens opacities and cataract and exudate form the back and side.
- Pellagra is a multiple deficiency disease, therefore, pellagrins develop symptoms of riboflavin deficiency.

Sources: Riboflavin is widely distributed throughout the plant and animal kingdoms. Milk, liver, kidney and heart are excellent sources, many vegetables are also very good sources.

Vitamin B_4 and B_5 Niacin (Nicotinic Acid)

- An important disease known as pellagra, which was

responsible for thousands of lives annually was caused due to the deficiency of nicotinic acid.

- People suffering from pellagra develop digestive disturbances characterised by diarrhea, fiery red tongue, ulcers in the mouth, dermatitis loss of appetite and other symptoms.
- In pigs, loss of weight, diarrhea, vomiting, dermatitis and normocytic anaemia are common symptoms.
- Nicotinic acid is also a dietary essential for adult fouel,
- In chickens its deficiency is accompained by poor growth, mouth symptoms, somewhat similar to that of black tongue in dogs.

Sources: Cereals and its by-products except corn are good sources. Groundnut oil is an excellent source. Alfalfa and other leafy material are fair sources, milk, coffee, yeasts are also good sources.

Vitamin B_6 (Pyridoxine)

- The vitamin was isolated by Keristezy (1938) and the synthesis of the vitamin was accomplished by Harris and Folkers in U.S.

Deficiency Symptoms (disease): In chicks, a deficiency causes acute convulsion, flatter on pan, usually starts kicking and generally die.

- In adult birds with mild deficiency hatchability and egg production are reduced.
- In rat characteristic skin lesion appear in the peripheral parts of the body such as paws, nose, ears, tails, etc.

Sources: Good sources of vitamin include yeast and certain seeds, such as wheat and maize liver and green leafy vegetables.

Biotin (Vitamin H)

- Biotin is a relatively simple monocarboxylic acid, biotin is very slightly soluble in water and alcohol.

Deficiency Disease: In chicks biotin deficiency results in dermatitis similar to that occurring in pantothenic acid deficiency.

- By the deficiency of biotin various disease occurred like muscle pains, slight anaemia, anorexia, insomnia, etc.

Sources: Richest sources of biotin are royal jelly, liver, yeast, molasses peanuts and eggs most fresh vegetables are fairly good sources.

Vitamin B_{12} (Cyanocobalamin)

- Vitamin B_{12} has been isolated in several different biologically active forms.
- Vitamin B_{12} anti-anaemia factor may play an important role in preventing and curing pernicious anaemia.
- It is a growth factor for chickens, rats, turkey and pigs, children are also affected in case of rat. Studies have shown that under certain conditions the vitamin is required for reproduction lactation as well as for growth.
- It is believed to be synthesised in rumen by sheep and cattle.
- Liver is the richest sources (natural sources).
- Alfalfa, pasture grass, etc. also very good source of vitamin B_{12}.

Characteristics of Cattle Breeds

- Cattle belong to:

Phylum	—	Chordata
Class	—	Mammalia
Order	—	Artiodactyla
Family	—	Brovidae

Division of Indian Breeds of Cattle

(a) Classification of cattle breeds on the basis of types.

(b) Classification of cattle breeds on the basis of utility.

Indian Cattle Breeds

Milch breeds	*Dual purpose breeds*	*Draught-purpose breeds*
1. Gir	1. Haryana	1. Hallikar
2. Sahiwal	2. Ongole	2. Amritmahal
3. Red Sindhi	3. Kankrej	3. Khillari
4. Tharparkar	4. Deoni	4. Bargur
	5. Nimari	5. Nagori
	6. Dangi	6. Malvi
	7. Rathi	7. Ponwar
		8. Siri
		9. Gaolao
		10. kenkatha
		11. kangayam
		12. Krishan valley

Milch Nature: Breeds

1. Gir

- The origin of gir breed is gir forest of South Kathiawar.

Characteristics: Well proportioned body, ears are markedly long, the head is moderatly long but massive in appearance with prominent bony forehead, colour is seldom entire varying from almost red to almost black, spot of different colours are one the chief characteristics of this breeds. Average female weight is 386 kg while males weigh 544 kg.

Production: Cows are good milkers. Average yield 1746 kg. Bullocks are heavy and powerful.

2. ***Sahiwal***

- The origin of this breed is central and southern dry areas of the Punjab particularly in Montgomery district in Pakistan.

Characteristics: Deep body, loose skin, short legs, stumpy horns, broad head and lethargic horns are short and thick. Various sheds of red, pale red and dark brown splashed with white colour.

Production: This is the best dairy breeds in India. In village the average yield is 2150 kg in 300 days lactation.

3. ***Red Sindhi***

- The home of this breed is round about Karanchi and Hyderabad.

Characteristics: Medium size and compact, animals having well proportioned body, extremely docile, deep dark colour, heavy hump develop and sheath. Male weight average 450 kg while females 295 kg.

Production: Yield as high as 5,433 kg per lactation. Average yield 1.474 kg. Bullocks suited for field works.

4. ***Tharparkar***

- The home of the breed is the Tharparkar district of Hyderabad.

Distinguishing Characteristics: Medium size, deep built, short, straight and strong limbs. Strong and well proportioned frame, broad poll and forehead is slightly convex with medium sized horns. The tail is fine with black switch.

Production: Cows are good yielders, bullocks suited for carting and ploughing. Average yield in village condition is 1474 kg.

Dual Target of Breeds

1. *Haryana*

Origin: The Haryana breed originated in east Punjab and are now extensively found in Rohtak, Hisar, Gurgoan, etc.

Features: Proportionate body, head is carried high. Horns are short, curving upward. Popular colour is white or light grey. Long and narrow face, small and sharp ears, skin is of fine texture and close to the body. Legs are moderately long and lean. Tail is short, thin and tapering towards the end with a black switch reaching just below the hocks.

Production: Bullocks are good working animals for fast ploughing and road transport. Cow are average milkers with 1400 kg/lactation.

2. *Ongole*

Origin: Ongole tract of Andhra Pradesh.

Characteristics: Large, heavy, muscular, forehead is broad with stumpy. Horns thick at the base and firm without cracks. Hump is well developed and erect, filled up on both sides without leaning. Popular colour is white.

Production: Govt. Live-stock Research Farm Guntur has maintained this breed. Bullocks are powerful and suitable for cart and road work but are not fast cow are good yielder.

3. *Kankerj*

Origin: This breed originated in north Gujarat in the Mumbai province of India. Chief Characteristics: This is the heaviest breed, broad chest, forehead dished in the centre. Hump is well developed, tough skin, powerful body.

- In females colour markings are lighter.
- Tail is of moderate length with black switch.

- The animal is energetic and vigorous, excitable and nervous with strangers.

Production: Cows are fairly good milkers. Average yield is about 1333 kg but individual approach 3500 kg.

4. Deoni

Origin: North-West and western portion of Hyderabad.

Chief Characteristics: Resembles gir breed, less pronounced forehead. Colour is black and white or red and white with irregular patches. Deep chest, well arched ribs.

Production: Cow are good milker. Average yield about 900 kg in 300 days. Bullocks are well suited for heavy work.

Drought Nature of Breeds

Amritmahal

Origin: Karnataka State.

Cheif Characteristics: Compact form with short straight back, well arched ribs, narrow face and prominent forehead with a furrow in the middle long sweeping horns typical of Karnataka a type cattle. Tail is of moderate length with a black switch grey colour body.

Production: This breed is the best draught breeds of India. Cow are not so good milkers.

Kanagayam

Origin : This name of breed derived from Kanagayam division of Dharampuram Taluk of Coimbatore district in Tamil Nadu.

Chief Characteristics: Strong horns with sharp tips. Body moderatly long, straight back, short and strong neck. The colour of the bull is of grey with dark gray to black marking, while cow is white with black markings. Just in front of the fetlocks on all four legs, moderates sized hump, strong limbs.

Yield: The cow are poor milkers yielding about 666 kg per lactation. The bulls are excellent type for hard work.

Malvi

- This breed are found in malwa district in M.P. but now also found in Rajasthan.

Special Qualities: Deep short and compact body. Ears are alert and short in size, short but thick horn tapering to a blunt point, short neck. Tail is moderately longwith black switch. The general colour of the breed is grey to iron grey.

Yield: Bullocks are good for road and field work, good adaptability, cows are poor milkers 5-6 kg/day. This breed is essentially a good draught breed.

Siri

- These animals are found in hill part of Darjeeling, Sikkim and Bhutan.

Chief Characters: Massive body, small head, sharp horns, small ears, well placed hump covered with a tuft of hair at the top. Strong legs, colour is black and white or red and white.

Yield: Bullocks are well suited for cart purposes cows are poor milkers.

Hallikar

- This breed is mainly found in the district of Hassan in Karnataka state.

Chief Characters: This is the medium size animal compact and muscular in appearance. The head is usually long. The face is long with small ears. The hump is moderately developed.

Production: Cows are poor milkers and bullock are excellent draught type suitable both for road and field work.

New Breeds: Emergence

Karan Swiss

- This breed has been evolved at the NDRI, Karnal in Haryana by the breeding the Sahiwal cows with the frozen semen of brown Swiss bull.
- Cow body weight is 400-500 kg.
- Bully body weight is 600-750 kg.
- Ears are small.
- Neck is of medium size.
- Hump is long. The colour is light grey to deep brown.

Production: The lactation yield is 3355 kg in 305 days. The highest daily yield recorded is 43 kg.

Sunandini

- This breed is originated in Kerala by crossing the local non-discript cattle with Jersy, brown Swiss.
- The overall milk production per lactation is more than 2500 litres.

Karan Fries

- This breed also originated at National Dairy Research Institute, Karnal.

Use of Domestic Buffalo

- The domestic or water buffalo *(Bubalus bubalis)* belongs to the genus Bubalus.
- Now a day the buffalo is commonly known as an "Asian Animal".
- The domesticated buffaloes may be classified into two main categories: 1. The swamp buffalo; 2. The river buffalo

The Swamp Buffalo: The swamp type buffalo has an inherent partiality for water and love wallowing in water or mud pool.

The River Buffalo: The river type buffalo show preference for clean running water.

- The total world buffaloes population of 138.37 million is less then 10% of the total world bovine population.
- The annual rate of increase in buffalo population over the last 10 years was 1.7% compared to 0.7% in cattle.
- Largest concentration of buffaloes is in the Asia and Pacific countries.

Wallowing: Its Meaning

- Wallowing means rolling or floundering in mud or in water. The river breeds prefer the clean water of streams and pools punching close together.
- But the swamp buffalo of south-east Asia like to wallow in slime, in a mudhole accomodating one animal.

Indian Buffaloes Breeds

Murrah group	*Gujarat group*	*Uttar Pradesh group*	*Central India group*	*South group*
1. Murrah	1. Surti	1. Bhadawari	1. Nagpuri	1. Toda
2. Nili Ravi	2. Jaffarbadi	2. Tarai	2. Pandhe-puri	2. South Kanara
3. Kundi	3. Mehsana		3. Manda	
4. Godavari			4. Jerangi	
			5. Kalhandi	
			6. Sambalpur	

- Buffaloes apparently wallow for two main reasons
  1. To regulate body temperature and
  2. To control parasite.
- Buffaloes are classified on the basis of regions

Description

Murrah Group's Breed

Murrah

- The origin of this breed is mainly in Delhi and Punjab.
- Popular colour is jet black with white markings on the tail, face and extremities.

Production: The average milking capacity in well established herds is about 1400 to 2000 kg and butter fat content of 7% of milk in a lactation of 9 to 10 months.

Nili Ravi

- This type of breeds are found in the valley of Montgomery and Ferozepur.
- This animal is medium sized, deep frame, horns are small with high coil and the neck is long, thin and fine.

Production: She-buffaloes are high milkers average lactation period of 250 days. Males are used for draught work.

Breed of Gujarat Group

Surti

- This breed is found in the south-western part of Gujarat state of India.
- The skin colour is black or brown and the colour of the hair varies from rusty brown to silver grey. The tail is long, this and flexible usually with a white switch.

Production: The average lactation yield of well-breed animals is 1600 kg with the fat content of 7.5%.

Jaffarbadi

- These breeds are found mainly in the gir forest of Kathiawar.
- Body is longer. Females are somewhat loose.
- Animal of this breed are generally black in colour. The Jaffarbadi males on the average weigh 600 kg. While females weigh 460 kg.

Production: Animals are heavy milkers ranging from 15 to 18 kg/day.

Mehsana

- The name of Mehsana derived from its home tract of the town Mehsana located in the north of the Gujarat state.
- Mehsana breed of buffalo was developed by cross of Murrah and Surti breed.
- The breed is a medium sized animal with a low-set deep body.

Breed Uttar Pradesh Group

Bhadawari

- These breeds are found in Bhadawari estate of Agra district and adjoining areas of Gwalior and Etawah. This is found scattered in the surrounding of Yamuna and Chambal rivers.
- Medium sized and wedge-shaped body. Legs are short but stout. Colour is copper, hair scanty. Eyes are prominent, active and bright.

Production: Average yield from 2000 to 2070 kg in lactation period of 305 days with a high fat percentage. Males are used for drought purpose.

Tarai

- The name of this breed derived from the Tarai area of U.P. The breed is native to hilly area.
- Its moderate body, horns are long and flat with coils. The colour of the skin varies from black to brown.

Production: The breed is poor milk producer, which may be as low as 2-3 kg daily. Males are used for draught purposes.

Breed of Central India Group

Nagpuri

- These animals are mainly found in Nagpur in Maharashtra.
- Colour is usually black; occasionally white markings are observed on the face, legs and switch. Neck is also longer with heavy brisket.

Production: The female are moderately good yielders, producing on an average 1000 litres milk/lactation of 300 days with butter content between 7.0-8.5%.

Pandhepuri

- This breed is broadly found in south Maharashtra, parts of A.P.
- Animals are of medium size with long narrow face and very long flat and usually twisted with horns.

Production: Males are hardy and well suited for drought purpose.

Manda

- This breed are found in the hills above Parlakimedi and Mandasa on the borders of Orissa and A.P.
- General colour of this breed is brown or grey with yellowish tuft of hair on the knees and fetlocks. The breed is medium sized animal. Eyes are sharp, horns are broad.

Production: Milk yield is satisfactory and males are hardy and can work in the hot sun.

Sambalpur

- The origin of the breed is Sambalpur area of Orissa.
- Animals are large and powerful having long narrow barrel, body colour is generally black.

Production: Females breed produce milk regularly and satisfactory. Males are active and good prefer for draught purpose.

Kalhandi

- The home of this breed is eastern part of A.P. and adjoining area of Orissa.
- Body colour of the animal is grey and ash grey colour. Eyes are prominent and large with narrow red margin around the lids. Tail is of medium length with white switch.

Production: Animals are hardy and docile. Milk yield of females is satisfactory. Males are used to carry load in hilly areas as well as for ploughing in plains.

Jerangi

- The breed is widely distributed in Jerangi hills of Orissa.
- This is the one of the dwarf breeds of buffalo and its height does not exceed four feet. The body colour is black.

Production: Females are poor for milk production but males are strictly used for ploughing in water logged paddy field.

Breed of South India Group

1. *Toda:* This breed is found in the Nilgiri hills of Tamil Nadu state.

2. Buffaloes the cows which become still weaker during the hot dry month (April-June). It is known as silent heat.
3. High buffalo calf mortality is one of the serious problem among dairy farmers. Various disease are responsible for more calf matality during winter season like pneumonia and II highest cause of calf is gastroenteritis. This disease may be due to viral, bacterial or parasitic infection. So good care are essential for devoid of this problem.
4. Services are given before feeding, as bulls seem to perform more quickly if this procedure is followed. If the interval between two successive, services is long the bull is allowed to serve a second time.
5. Buffalo bull attains sexual maturity later than ox bull.
6. The buffaloes are less distinguish in foraging and therefore consume a larger quantity of coarse fodder that is not readily eaten by cattle. This might be one of the reason why buffaloes thrive better than do cattle on coarse fodder.
7. During hot summer months, buffaloes need special comfort for which they love to have wallowing.
8. Buffalo bulls are more strong than the cow bulls. Fights between buffalo bulls are dangerous end in fatality.
9. Buffaloes are more susceptible to some disease like rinderpest, haemorrhagic septicaemia than the indigenous cattle managed under comparable conditions.
10. India possesses the largest number of buffaloes (84.21 million in 1992). India is followed by China, Pakistan, Thailand, Philippines, Nepal, Indonesia, Vietnam, etc.
11. Buffaloes have a low cost of milk production per litre which is a net outcomes of physiological superiority of buffaloes over cows.

- There are large-sized animals having long barrel and strong build. Horns of these buffaloes are variable in shape. Face is short and wide. The animals are not always docile.

Production: Females are good milkers, yield between 4.5 to 8.5 litres daily with an average of 7.0% fat.

Buffalo Milk

- Indian buffaloes gives more milk than the Indian cow. Buffaloes have been contributing over 56% of total milk production. It has more total solids and the milk is richer in calcium and phosphorus and lower in sodium and potassium.
- Buffalo milk and ghee is white while cow milk and ghee is golden yellow in colour due to conversion of carotinoids into vitamin A.
- Buffalo milk is less suitable than cows milk for making Ghee and rosogulla.
- Breeding season of buffaloes is November to February because during this period bulls have been found to be very active sexually and quality and quantity of semen is very high and she buffaloes show maximum ovarian activity.
- The estrus cycle is about 21 days which divided into 3 phases: (i) Pro-estrus—This may continue for 2-3 days; (ii) Meta-estrus—The period of meta-estrus is 3-4 days; (iii) Dio-estrus—This is the Last phase of estrous cycle 10-18 days.

Suitable Information About Buffalo

1. Mostly she-buffaloes show sign of heat at night (mid night).
2. The oestrus symptoms is weaker in.

Natural Resources

Life on this planet earth depends upon a large number of things and services provided by the nature, which are known as Natural resources. Thus water, air, soil, minerals, coal, forests, crops and wildlife are all examples of natural resources.

The natural resources are of two kinds:

- Renewable resources which are in-exhaustive and can be regenerated within a given span of time e.g. forests, wildlife, wind energy, Bio-mass energy, tidal energy, hydro power, etc. Solar energy is also a renewable form of energy as it is an inexhaustible source of energy.
- Non-renewable resources which cannot be regenerated e.g. Fossil fuels like coal, petroleum, minerals, etc. Once we exhaust these reserves, the same cannot be replenished.

Even our renewable resources can become non-renewable if we exploit them to such extent that their rate of consumption exceeds their rate of regeneration. For example, if a species is exploited so much that its population size declines below the threshold level then it is not able to sustain itself and gradually the species becomes endangered or extinct. It is very important to protect and conserve our natural resources and use them in a judicious manner so that we don't exhaust them. It does not mean that we should stop using most of the natural resources. Rather, we should use the resources in such a way that we always save enough of them for our future generations. In this unit we shall discuss the major natural resources: (i) Forest resources, (ii) Water resources, (iii) Mineral resources, (iv) Food resources, (v) Energy resources, (vi) Land resources.

Various Means: Forest

Forests are one of the most important natural resources on this earth. Covering the earth like a green blanket these forests not only produce innumerable material goods, but also provide several environmental services which are essential for life.

About l/3rd of the world's land area is forested which includes closed as well as open forests. Former USSR accounts for about a 5th of the world's forests, Brazil for about a 7th and Canada and USA each for 6-7%. But it is a matter of concern that almost everywhere the cover of the natural forests has declined over the years. The greatest loss occurred in tropical Asia where one third of the forest resources have been destroyed.

Benefits of Jungles

Commercial uses: Forests provide us a large number of commercial goods which include timber, firewood, pulpwood, food items, gum, resins, non-edible oils, rubber, fibers, lac, bamboo canes, fodder, medicine, drugs and many more items, the total worth of which is estimated to be more than $ 300 billion per year.

Half of the timber cut each year is used as fuel for heating and cooking. One third of the wood harvest is used for building materials as lumber, plywood and hardwood, particle board and chipboard. One sixth of the wood harvest is converted into pulp and used for paper industry. Many forest lands are used for mining, agriculture, grazing, and recreation and for development of dams.

Ecological uses: While a typical tree produces commercial goods worth about $ 590 it provides environmental services worth nearly $ 196, 250. The ecological services provided by our forests may be summed up as follows:

- *Production of Oxygen:* The trees produce oxygen by photo synthesis which is so vital for life on this earth. They are rightly called as earth's lungs.

- *Reducing Global Warming:* The main greenhouse gas carbon dioxide (CO_2) is absorbed by the forests as a raw material for photosynthesis. Thus forest canopy acts as a sink for CO_2 thereby reducing the problem of global warming caused by greenhouse gas CO_2.
- *Wildlife Habitat:* Forests are the homes of millions of wild animals and plants. About 7 million species are found in the tropical forests alone.
- *Regulation of Hydrological Cycle:* Forested watersheds act like giant sponges, absorbing the rainfall, slowing down the runoff and slowly releasing the water for recharge of springs. About 50-80 %of the moisture in the air above tropical forests comes from their transpiration which helps in bringing rains.
- *Soil Conservation:* Forests bind the soil particles tightly in their roots and prevent soil erosion. They also act as wind breaks.
- *Pollution Moderators:* Forests can absorb many toxic gases and can help in keeping the air pure. They have also been reported to absorb noise and thus help in preventing air and noise pollution.

Since time immemorial, humans have depended heavily on forests for food, medicine, shelter, wood and fuel. With growing civilisation the demands for raw material like timber, pulp, minerals, fuel wood, etc. shooted up resulting in large scale logging, mining, road-building and clearing of forests.

Our forests contribute substantially to the national economy. The international timber trade alone is worth over US $ 40 billion per year. Excessive use of fuel wood and charcoal, expansion of urban, agricultural and industrial areas and overgrazing have together led to over-exploitation of our forests leading to their rapid degradation.

Deforestation: The total forest area of the world in 1900 was estimated to be 7,000 million hectares which was reduced to 2890 million ha in 1975 and fell down to just 2,300 million ha by 2000. Deforestation rate is relatively less in temperate countries, but it is very alarming in tropical countries where it is as high as 40-50 per cent and at the present rate it is estimated that in the next 60 years we would lose more than 90 per cent of our tropical forests.

The forested area in India seems to have stabilised since 1982 with about 0.04% decline annually between 1982-90. FAO (1983) estimated that about 1.44 m ha of land was brought under afforestation during this period leading to stabilisation. As per FAO estimates, the deforestation rate per unit population in India is the lowest amongst the major tropical countries, despite the fact that we have a huge population size and very low per capita forest area (0.075 ha per capita). However, we are still far behind the target of achieving 33% forest area, as per our National Forest Policy, as we are still having only 19.27 % of our land area (63.38m ha), covered by forests based on satellite data (MoEF, 1998)

Principal Reasons behind Deforestation

(i) *Shifting Cultivation:* There are an estimated 300 million people living as shifting cultivators who practice slash and burn agriculture and are supposed to clear more than 5 lakh hectares of forests for shifting cultivation annually. In India, we have this practice in North-East and to some extent in Andhra Pradesh, Bihar and M.P. which contribute to nearly half of the forest clearing annually.

(ii) *Fuel Requirements:* Increasing demands for fuel wood by the growing population in India alone has shooted up to 300-500 million tons in 2001 as compared to just 65 million tons during independence, thereby increasing the pressure on forests.

(iii) *Raw Materials for Industrial Use:* Wood for making boxes, furniture, railway-sleepers, plywood, match-boxes, pulp for paper industry, etc. have exerted tremendous pressure on forests. Plywood is in great demand for packing tea for Tea industry of Assam while fir tree wood is exploited greatly for packing apples in J & K.

(iv) *Development Projects:* Massive destruction of forests occur for various development projects like hydroelectric projects, big dams, road construction, mining, etc.

(v) *Growing Food Needs:* In developing countries this is the main reason for deforestation. To meet the demands of rapidly growing population, agricultural lands and settlements are created permanently by clearing forests.

(vi) *Overgrazing:* The poor in the tropics mainly rely on wood as a source of fuel leading to loss of tree cover and the cleared lands are turned into the grazing lands. Overgrazing by the cattle leads to further degradation of these lands.

Principal Results of Deforestation: Deforestation has far reaching consequences, which may be outlined as follows:

(i) It threatens the existence of many wildlife species due to destruction of their natural habitat.

(ii) Biodiversity is lost and along with that genetic diversity is eroded.

(iii) Hydrological cycle gets affected, thereby influencing rainfall.

(iv) Problems of soil erosion and loss of soil fertility increase.

(v) In hilly areas it often leads to landslides.

Tasks of Forests

Timber Extraction: Logging for valuable timber, such as teak and Mahogany not only involves a few large trees per hectare but about a dozen more trees since they are strongly interlocked with each other by vines, etc. Also road construction for making approach to the trees causes further damage to the forests.

Mining: Mining operations for extracting minerals and fossil fuels like coal often involves vast forest areas. Mining from shallow deposits is done by surface mining while that from deep deposits is done by sub-surface mining. More than 80,000 ha of land of the country is presently under the stress of mining activities. Mining and its associated activities require removal of vegetation along with underlying soil mantle and overlying rock masses. This results in defacing the topography and destruction of the landscape in the area.

Large scale deforestation has been reported in Mussorie and Dehradun valley due to indiscriminate mining of various minerals over a length of about 40 Km. The forested area has declined at an average rate of 33% and the increase in non-forest area due to mining activities has resulted in relatively unstable zones leading to landslides.

Indiscriminate mining in forests of Goa since 1961 has destroyed more than 50,000 ha of forest land. Coal mining in Jharia, Raniganj and Singrauli areas have caused extensive deforestation in Jharkhand. Mining of magnesite and soap-stones have destroyed 14 ha of forest in the hill slopes at Khirakot, Kosi valley, Almora.

Mining of radioactive minerals in Kerala, Tamil Nadu and Karnataka are posing similar threats of deforestation. The rich forests of Western Ghats are also facing the same threat due to mining projects for excavation of copper, chromite, bauxite and magnetite.

How Dams have its Imparts on Forests and People ?: Big dams and river valley projects have multi-purpose uses and have been referred to as *"Temples of modern India"*. However, these dams are also responsible for the destruction of vast areas of forests. India has more than 1550 large dams, the maximum being in the state of Maharashtra (more then 600), followed by Gujarat (more then 250) and Madhya Pradesh (130). The highest one is *Tehri dam,* on river Bhagirathi in Uttaranchal and the largest in terms of capacity is Bhakra dam on river Satluj in H.P. Big dams have been in sharp focus of various environmental groups all over the world which is mainly because of several ecological problems including deforestation and socio-economic problems related to tribal or native people associated with them. The *Silent Valley* hydroelectric project was one of the first such projects situated in the tropical rain forest area of Western Ghats which attracted much concern of the people. *The crusade against the ecological damage and deforestation caused due to Tehri dam was led by Sh. Sunder Lal Bahuguna, the leader of Chipko movement. The cause of Sardar Sarovar Dam related issues has been taken up by the environmental activists Medha Patekar, joined by Arundhati Roy and Baba Amte.*

For building big dams, large scale devastation of forests takes place which breaks the natural ecological balance of the region. Floods, droughts and landslides become more prevalent in such areas. Forests are the repositories of invaluable gifts of nature in the form of biodiversity and by destroying them (particularly, the tropical rain forests) we are going to lose these species even before knowing them. These species could be having marvelous economic or medicinal value and deforestation results in loss of this storehouse of species which have evolved over millions of years in a single stroke.

Means of Water

Water is an indispensable natural resource on this earth on which all life depends. About 97% of the earth's surface is covered by water and most of the animals and plants have 60-65% water in their body.

Water is characterised by certain unique features which make it a marvellous resource:

(i) It exists as a liquid over a wide range of temperature, i.e. from 0° to 100°C.

(ii) It has the highest specific heat, due to which it warms up and cools down very slowly without causing shocks of temperature jerks to the aquatic life.

(iii) It has a high latent heat of vaporisation. Hence, it takes a huge amount of energy for getting vaporised. That's why it produces a cooling effect as it evaporates.

(iv) It is an excellent solvent for several nutrients. Thus, it can serve as a very good carrier of nutrients, including oxygen, which are essential for life. But, it can also easily dissolve various pollutants and become a carrier of pathogenic microorganisms.

(v) Due to high surface tension and cohesion it can easily rise through great heights through the trunk even in the tallest of the trees like Sequoia.

(vi) It has an anomalous expansion behaviour, i.e. as it freezes, it expands instead of contracting and thus becomes lighter. It is because of this property that even in extreme cold, the lakes freeze only on the surface. Being lighter the ice keeps floating, whereas the bottom waters remain at a higher temperature and therefore, can sustain aquatic organisms even in extreme cold.

The water we use keeps on cycling endlessly through the environment, which we call as Hydrological Cycle. We have enormous resources of water on the earth amounting to about 1404 million Km^3. The water from various moist surfaces evaporates and falls again on the earth in the form of rain or snow and passes through living organisms and ultimately returns to the oceans.

Every year about 1.4 inch thick layer of water evaporates from the oceans, more than 90% of which returns to the oceans through the hydrological cycle. Solar energy drives the water cycle by evaporating it from various water bodies, which subsequently return through rainfall or snow. Plants too play a very important role by absorbing the groundwater from the soil and releasing it into the atmosphere by the process of transpiration.

Global distribution of water resources is quite uneven depending upon several geographic factors. Tropical rain forest areas receive maximum rainfall while the major world deserts occur in zones of dry, descending air (20-40° N and S) and receive very little rainfall.

Use and Misuse of Water: Due to its unique properties water is of multiple uses for all living organisms. Water is absolutely essential for life. Most of the life processes take place in water contained in the body. Uptake of nutrients, their distribution in the body, regulation of temperature, and removal of wastes are all mediated through water.

Human beings depend on water for almost every developmental activity. Water is used for drinking, irrigation, transportation, washing and waste disposal for industries and used as a coolant for thermal power plants. Water shapes the earth's surface and regulates our climate.

Water Use by Humans is of Two Types:

Water Withdrawal: taking water from groundwater or surface water resource and water consumption: the water which is taken up but not returned for reuse. Globally, only about 60 per cent of the water withdrawn is consumed due to loss through evaporation.

With increasing human population and rapid development, the world water withdrawal demands have increased many folds and a large proportion of the water withdrawn is polluted due to anthropogenic activities. On a global average 7.0 per cent of the water withdrawn is used for agriculture. In India, we use 93% of water in agricultural sector while in a country like Kuwait, which is water-poor, only 4% is used for watering the crops. About 25% of water on global average is used in industry, which again varies from a high of 70% in European countries to as low as 5% in less developed countries. Per capita use of water shows wide variations. In USA, an average family of 4 consumes more than 1000 M^3 of water per year, which is many times more than that in most developing countries.

Precious Means of Nature: Although water is very abundant on this earth, yet it is very precious. Out of the total water reserves of the world, about 97% is salty water (marine) and only 3% is fresh water. Even this small fraction of fresh water is not available to us as most of it is locked up in polar ice caps and just 0.003% is readily available to us in the form of groundwater and surface water.

Overuse of groundwater for drinking, irrigation and domestic purposes has resulted in rapid depletion of groundwater in various regions leading to lowering of water table and drying of wells. Pollution of many of the groundwater aquifers has made many of these wells unfit for consumption.

Rivers and streams have long been used for discharging the wastes. Most of the civilisations have grown and flourished on the banks of rivers, but unfortunately, growth in turn, has been responsible for pollution of the rivers.

As per the United Nations estimates (2002), at least 101 billion people do not even have access to safe drinking water and 2.4 billion do not have adequate sanitation facilities. Increasing population and expanding development would further increase the demands for wastes. It is estimated that by 2024, two-thirds of the world population would be suffering from acute water shortage.

Groundwater: About 9.86% of the total fresh water resources is in the form of groundwater and it is about 35-50 times that of surface water supplies. Till some time back groundwater was considered to be very pure. However, of late, even groundwater aquifers have been found to be contaminated by leachates from sanitary landfills, etc.

A layer of sediment or rock that is highly permeable and contains water is called an aquifer. Layers of sand and gravel are good aquifers while clay and crystalline rocks (like granite) are not since they have low permeability. Aquifers may be of two types: Unconfined aquifers which are overlaid by permeable earth materials and they are recharged by water seeping down from above in the form of rainfall and snow melt.

Confined aquifers which are sand witched between two impermeable layers of rock or sediments and are recharged only in those areas where the aquifer intersects the land surface. Sometimes the recharged area is hundreds of kilometers away from the location of the well. Groundwater is not static, it moves, though at a very slow rate of about a meter or so in a year.

Influences of Groundwater

(i) *Subsidence:* When groundwater withdrawal is more than its recharge rate, the sediments in the aquifer get compacted, a phenomenon known as *ground subsidence.* Huge economic losses may occur due to this phenomenon because it results in the sinking of overlying land surface. The common problems associated with it include structural damage in buildings, fracture in pipes, reversing the flow of sewers and canals and tidal flooding.

(ii) *Lowering of Water Table:* Mining of groundwater is done extensively in arid and semi-arid regions for irrigating crop fields. However, it is not advisable to do excessive mining as it would cause a sharp decline in future agricultural production, due to lowering of water table.

(iii) *Water Logging:* When excessive irrigation is done with brackish water it raises the water table gradually leading to water-logging and salinity problems.

Surface Water: The water coming through precipitation (rainfall, snow) when does not percolate down into the ground or does not return to the atmosphere as evaporation or transpiration loss, assumes the form of streams, lakes, ponds, wetlands or artificial reservoirs known as surface water. The surface water is largely used for irrigation, industrial use, public water supply, navigation, etc. A country's economy is largely dependent upon its rivers.

Countries with Rich and Poor Water Resources: The top ten water rich countries are Iceland, Surinam, Guyana, Papua New Guinea, Gabon, Solomon Islands, Canada, Norway, Panama, and Brazil lying in the far north and have low evaporation losses.

The water poor countries include Kuwait, Egypt, United Arab Emirates, Malta, Jordon, Saudi Arabia, Singapore, Maldovia, Israel and Oman, lying in the desert belt at about 15° to 25° Latitude and some of them like Malta and Singapore are densely populated areas resulting in low per capita water.

Floods: In some countries like India and Bangladesh rainfall does not occur throughout the year, rather, 90% of it is concentrated into a few months (June-September). Heavy rainfall often causes floods in the low-lying coastal areas. Prolonged downpour can also cause the over-flowing of lakes and rivers resulting into floods. Deforestation, overgrazing, mining, rapid industrialisation, global warming, etc. have also contributed largely to a sharp rise in the incidence of floods, which otherwise is a natural disaster.

Floods have been regular features of some parts of India and Bangladesh causing huge economic loss as well as loss of life. People of Bangladesh are accustomed to moderate flooding during monsoon and they utilise the flood water for raising paddy. But, severe floods like that in 1970, 1988 and 1991 resulting from excessive Himalayan runoff and storms, had very disastrous consequences causing massive deaths and damages. In 1970, about one million people were drowned while 1,40,000 people died in 1991. Networking of rivers is being proposed at national level to deal with the problems of floods.

Droughts: There are about 80 countries in the world, lying in the arid and semi-arid regions that experience frequent spells of droughts, very often extending up to year long duration. When annual rainfall is below normal and less than evaporation, drought conditions are created.

Ironically, these drought-hit areas are often having a high population growth which leads to poor land use and makes the situation worse.

Anthropogenic Causes: Drought is a meteorological phenomenon, but due to several anthropogenic causes like over grazing, deforestation, mining, etc. there is spreading of the deserts tending to convert more areas to drought affected areas. In the last twenty years, India has experienced more and more desertification, thereby increasing the vulnerability of larger parts of the country to droughts.

Erroneous and intensive cropping pattern and increased exploitation of scarce water resources through well or canal irrigation to get high productivity has converted drought-prone areas into decertified ones. In Maharashtra there has been no recovery from drought for the last 30 years due to over-exploitation of water by sugarcane crop which has high water demands.

Remedial Measures: Indigenous knowledge in control of drought and desertification can be very useful for dealing with the problem. Carefully selected mixed cropping help optimise production and minimise the risks of crop failures. Social Forestry and Wasteland development can prove quite effective to fight the problem, but it should be based on proper understanding of ecological requirements and natural process, otherwise it may even boomerang. The Kolar district of Karnataka is one of the leaders in Social Forestry with World Bank Aid, but all its 11 talukas suffer from drought. It is because the tree used for plantation here was *Eucalyptus* which is now known to lower the water table because of its very high transpiration rate.

Conflicts of Water Sharing: Indispensability of water and its unequal distribution has often led to inter-state or international disputes. Issues related to sharing of river water have been largely affecting our farmers and also shaking our governments. Some major water conflicts are discussed here.

Water-dispute in the Middle East: Three river basins, namely the Jordan, the Tigris-Euphrates and the Nile are the shared water resources for Middle East countries. Ethiopia controls the head waters of 80% of Nile's flow and plans to increase it.

Sudan too is trying to divert more water. This would badly affect Egypt, which is a desert, except for a thin strip of irrigated cropland along the river Nile and its delta. The population of Egypt is likely to double in the next 20 years, thereby increasing its water crisis. Likewise there is a fierce battle for water among Jordan, Syria and Israel for the Jordan River water share.

Turkey has abundant water and plans to build 22 dams on Tigris-Euphrates for Hydroelectric power generation. But, it would drastically reduce the flow of water to Syria and Iraq, lying downstream. Turkey dreams to become the region's water Super power. It plans to transport and sell water to starved Saudi Arabia, Kuwait, Syria, Israel and Jordan. Probably, the next war in the Middle East would be fought over water and not oil.

The Indus Water Treaty: The Indus, one of the mightiest rivers is dying a slow death due to dams and barrages that have been built higher upon the river. The Sukkur barrage (1932), Ghulam Muhammad Barrage at Kotri (1958) and Tarbela and Chasma Dams on Jhelum, a tributary of Indus have resulted in severe shrinking of the Indus delta. In 1960, the Indus water treaty was established by which Indus, the Jhelum and the Chenab were allocated to Pakistan and the Satluj, the Ravi and the Beas were allocated to India. Being the riparian state, India has pre-emptive right to construct barrages across all these rivers in Indian territory. However, the treaty requires that the three rivers allocated to Pakistan may be used for *non-consumptive* purposes by India, i.e. without changing its flow and quality. With improving political relations between the two countries it is desirable to work out techno-economic details and go for an integrated development of the river basin in a sustainable manner.

The Cauvery Water Dispute: Out of India's 18 major rivers, 17 are shared between different states. In all these cases, there are intense conflicts over these resources which hardly seem to resolve. The Cauvery river water is a bone of contention between Tamil Nadu and Karnataka and the fighting is almost hundred years old. Tamil Nadu, occupying the downstream region of the river wants water-use regulated in the upstream. Whereas, the upstream state Karnataka refuses to do so and claims its primacy over the river as upstream user. The river water is almost fully utilised and both the states have increasing demands for agriculture and industry.

The consumption is more in Tamil Nadu than Karnataka where the catchment area is more rocky. On June 2,1990, the Cauvery Water Dispute Tribunal was set up which through an interim award directed Karnataka to ensure that 205 TMCF of water was made available in Tamil Nadu's Mettur dam every year, till a settlement was reached. In 1991-92 due to good monsoon, there was no dispute due to good stock of water in Mettur, but in 1995, the situation turned into a crisis due to delayed rains and an expert Committee was set up to look into the matter which found that there was a complex cropping pattern in Cauvery basin.

Sambra paddy in winter, *Kurvai* paddy in summer and some cash crops demanded intensive water, thus aggravating the water crisis. Proper selection of crop varieties, optimum use of water, better rationing, rational sharing patterns, and pricing of water are suggested as some measures to solve the problem.

The Canal Dispute of Satluj-Yamuna Link (SYL): The issue of sharing the Ravi-Beas waters and SYL issue between Punjab and Haryana is being discussed in the Supreme Court. The Eradi Tribunal (1985) based the allocation of water on the basis of the time-inflow data of 20 years (1960-80), according to which 17.17 MAF (million acre feet) water was available. However, now it is argued by Punjab that in the last 17 years there has been consistent decline reducing the quantity to 14.34 MAF. The Supreme Court on January 15, 2002 directed Punjab to complete and commission the SYL within a year, failing which the Centre was told to complete it. However, two years have passed, but neither the SYL has been completed nor the conflict over sharing of Ravi-Beas water is resolved.

The conflict is that Punjab being the riparian state for Beas, Ravi and Satluj stakes its claim, Haryana has faced acute shortage of water after it became a state in 1966 and has been trying to help it out by signing an MOU (Memorandum of understanding) with UP, Rajasthan and Delhi for allocation of Yamuna waters. The Yamuna basin covers the state of Haryana while the Indus basin covers Punjab. The conflict revolving around sharing of river water needs to be tackled with greater understanding and objectivity.

System of Traditional Water Management: In India, even today, there are several villages where water management is done not by the Irrigation Department, but by local managers. In south India, a *neerkatti* manages the traditional tanks very efficiently based on his/her knowledge of the terrain, drainage and irrigation needs. They usually give preference to the tail end fields and decide per capita allocation of water based on the stock of available water in the tank and crop needs. In Maharashtra, the water mangers are called *havaldars* or *jaghyas* who manage and resolve conflicts by overseeing the water channels from main canal to the distributary canals. In Laddakh, the water manager is known as *churpun* who has got complete charge with full powers over allocation of available water.

The major source of water is melt water from glaciers and snow supplementary by water from springs and marshes. The water is distributed to different fields through an intricate network of earthen channels. In traditional water management, innovative arrangements ensure equitable distribution of water, which are democratically implemented. The 'gram-sabhas' approve these plans publicly. While water disputes between states and nations often assume battle like situations, our traditional water managers in villages prove to be quite effective.

Uses and Harms of Big Dams

Benefits: River valley projects with big dams have usually been considered to play a key role in the development process due to their multiple uses. India has the distinction of having the largest number of river-valley projects. These dams are often regarded as a symbol of national development. The tribals living in the area pin big hopes on these projects as they aim at providing employment and raising the standard and quality of life. The dams have tremendous potential for economic upliftment and growth. They can help in checking floods and famines, generate electricity and reduce water and power shortage, provide irrigation water to lower areas, provide drinking water in remote areas and promote navigation, fishery, etc.

Problems of Atmosphere: The environmental impacts of big-dams are also too many due to which very often the big dams become a subject of controversy. The impacts can be at the upstream as well as downstream levels.

The upstream problems include the following:

(i) Displacement of tribal people

(ii) Loss of forests, flora and fauna

(iii) Changes in fisheries and the spawning grounds

(iv) Siltation and sedimentation of reservoirs

(v) Loss of non-forest land

(vi) Stagnation and waterlogging near reservoir

(vii) Breeding of vectors and spread of vector-borne diseases

(viii) Reservoir induced seismicity (RIS) causing earthquakes

(ix) Growth of aquatic weeds.

(x) Microclimatic changes.

The downstream impacts include the following:

(i) Water logging and salinity due to over irrigation

(ii) Micro-climatic changes

(iii) Reduced water flow and silt deposition in river

(iv) Flash floods

(v) Salt water intrusion at river mouth

(vi) Loss of land fertility along the river since the sediments carrying nutrients get deposited in the reservoir.

(vii) Outbreak of vector-borne diseases like malaria. Thus, although dams are built to serve the society with multiple uses, but it has several serious side-effects. That is why now there is a shift towards construction of small dams or mini-hydel projects.

Means of Minerals

Minerals are naturally occurring, inorganic, crystalline solids having a definite chemical composition and characteristic physical properties. There are thousands of minerals occurring in different parts of the world. However, most of the rocks, we see everyday are just composed of a few common minerals like quartz, feldspar, biotite, dolomite, calcite, laterite, etc. These minerals, in turn, are composed of some elements like silicon, oxygen, iron, magnesium, calcium, aluminium, etc.

Uses and Exploitation: Minerals find use in a large number of ways in everyday use in domestic, agricultural, industrial and commercial sectors and thus form a very important part of any nation's economy. The main uses of minerals are as follows:

(i) Development of industrial plants and machinery.

(ii) Generation of energy e.g. coal, lignite, uranium.

(iii) Construction, housing, settlements.

(iv) Defence equipments-weapons, armaments.

(v) Transportation means.

(vi) Communication-telephone wires, cables, electronic devices.

(vii) Medicinal system-particularly in Ayurvedic System.

(viii) Formation of alloys for various purposes (e.g. phosphorite).

(ix) Agriculture—as fertilizers, seed dressings and fungicides (e.g. zineb containing zinc, Maneb-containing manganese, etc.).

(x) Jewellery—e.g. Gold, silver, platinum, diamond.

Based on their properties, minerals are basically of two types:

(i) Non metallic minerals e.g. graphite, diamond, quartz, feldspar.

(ii) Metallic minerals e.g. Bauxite, laterite, haematite, etc.

Use of metals by human beings has been so extensive since the very beginning of human civilisation that two of the major epochs of human history are named after them as Bronze Age and Iron Age. The reserves of metals and the technical know-how to extract them have been the key elements in determining the economy and political power of nations. Out of the various metals, the one used in maximum quantity is Iron and steel (740 million metric tons annually) followed by manganese, copper, chromium, aluminium and Nickel.

Distribution and uses of some of the major metallic and non-metallic minerals are given in Tables.

Major Reserves and Important uses of Some of the Major Metals

Metal	*Major World Reserves*	*Major Uses*
Aluminium	Australia, Guinea, Jamaica	Packaging food items, transportation, utensils, electronics
Chromium	CIS, South Africa	For making high strength steel alloys, In textile/tanning industries
Copper	U.S.A., Canada, CIS, Chile, Zambia	Electric and electronic goods, building, construction, vessels
Iron	CIS, South America, Canada, U.S.A.	Heavy machinery, steel production transportation means
Lead	North America, U.S.A., CIS	Leaded gasoline, Car batteries, paints, ammunition
Manganese	South Africa, CIS, Brazil, Gabon	For making high strength, heat-resistant steel alloys
Platinum group	South Africa, CIS	Use in automobiles, catalytic converters, electronics, medical uses.
Gold	South Africa, CIS, Canada	Ornaments, medical use, electronic use, use in aerospace
Silver	Canada, South Africa, Mexico	Photography, electronics jewellery
Nickel	CIS, Canada, New Caledonia	Chemical industry, steel alloys

Major uses of Some Non-metallic Minerals

Non-metal Mineral	*Major Uses*
Silicate minerals	Sand and gravel for construction, bricks, paving, etc.
Limestone	Used for concrete, building stone, used in agriculture for neutralizing acid soils, used in cement industry
Gypsum	Used in plaster wall-board, in agriculture
Potash, phosphorite	Used as fertilizers
Sulphur pyrites	Used in medicine, car battery, industry.

It is evident from the Tables that the CIS countries (The Commonwealth of Independent States, *i.e.* 12 republics of former USSR), the United States of America, Canada, South Africa and Australia are having the major world reserves of most of the metallic minerals.

Due to huge mineral and energy resources, the USA became the richest and the most powerful nation in the world in even less than 200 years. Japan too needs a mention here, as there are virtually no metal reserves, coal, oil and timber resources in Japan and it is totally dependent on other countries for its resources. But, it has developed energy efficient technologies to upgrade these resources to high quality finished products to sustain its economy.

Minerals are sometimes classified as Critical and Strategic.

Critical minerals are essential for the economy of a nation e.g. iron, aluminium, copper, gold, etc.

Strategic minerals are those required for the defence of a country e.g. Manganese, cobalt, platinum, chromium, etc.

Few Principal Minerals of India

(a) Energy generating minerals:

Coal and Lignite: West Bengal, Jharkhand, Orissa, M.P., A.P. *Uranium* (Pitchblende or Uranite ore): Jharkhand, Andhra Pradesh (Nellore, Nalgonda), Meghalaya, Rajasthan (Ajmer).

(b) Other commercially used minerals:

Aluminium (Bauxite Ore): Jharkhand, West Bengal, Maharashtra, M.P., Tamil Nadu.

Iron (Haematite and Magnetite Ore): Jharkhand, Orissa, M.P., A.P., Tamil Nadu, Karnataka, Maharashtra and Goa.

Copper (Copper Pyrites): Rajasthan (Khetri), Bihar, Jharkhand, Karnataka, M.P., West Bengal, Andhra Pradesh and Uttaranchal.

Impacts of Mineral Extraction and Use on Environment : The issue related to the limits of the mineral resources in our earth's crust or in the ocean is not so significant. More important environmental concern arises from the impacts of extraction and processing of these minerals during mining, smelting, etc.

Indian Scenario: India is the producer of 84 minerals the annual value of which is about Rs. 50,000 crore. At least six major mines need a mention here which are known for causing severe problems:

(i) Jaduguda Uranium Mine, Jharkhand—exposing local people to radioactive hazards.

(ii) Jharia coal mines, Jharkhand—underground fire leading to land subsidence and forced displacement of people.

(iii) Sukinda chromite mines, Orissa—seeping of hexavalent chromium into river posing serious health hazard, Cr^{6+} being highly toxic and carcinogenic.

(iv) Kudremukh iron ore mine, Karnataka—causing river pollution and threat to biodiversity.

(v) East coast Bauxite mine, Orissa—Land encroachment and issue of rehabilitation unsettled.

(vi) North-Eastern Coal Fields, Assam—Very high sulphur contamination of groundwater.

Impacts of Mining: Mining is done to extract minerals (or fossil fuels) from deep deposits in soil by using sub-surface mining or from shallow deposits by surface mining. The former method is more destructive, dangerous and expensive including risks of occupational hazards and accidents.

Surface mining can make use of any of the following three types:

(a) *Open-pit Mining:* in which machines dig holes and remove the ores (e.g. copper, iron, gravel, limestone, sandstone, marble, granite).

(b) *Dredging:* in which chained buckets and draglines are used which scrap up the minerals from under-water mineral deposits.

(c) *Strip Mining:* in which the ore is stripped off by using bulldozers, power shovels and stripping wheels (e.g. phosphate rocks).

The environmental damage caused by mining activities are as follows:

(i) *De-vegetation and De-facing of Landscape:* The topsoil as well as the vegetation are removed from the mining area to get access to the deposit. While large scale deforestation or de-vegetation leads to several ecological losses as already discussed in the previous section, the landscape also gets badly affected. The huge quantities of debris and tailings along with big scars and disruptions spoil the aesthetic value of the region and make it prone to soil erosion.

(ii) *Subsidence of Land:* This is mainly associated with underground mining. Subsidence of mining areas often results in tilting of buildings, cracks in houses, buckling of roads, bending of rail tracks and leaking of gas from cracked pipelines leading to serious disasters.

(iii) *Groundwater Contamination:* Mining disturbs the natural hydrological processes and also pollutes the groundwater. Sulphur, usually present as an impurity in many ores is known to get converted into sulphuric acid through microbial action, thereby making the water acidic. Some heavy metals also get leached into the groundwater and contaminate it posing health hazards.

(iv) *Surface Water Pollution:* The acid mine drainage often contaminates the nearby streams and lakes. The acidic water is detrimental to many forms of aquatic life. Sometimes radioactive substances like uranium also contaminate the water bodies through mine wastes and kill aquatic animals. Heavy metal pollution of water bodies near the mining areas is a common feature creating health hazards.

(v) *Air Pollution:* In order to separate and purify the metal from other impurities in the ore, smelting is done which emits enormous quantities of air pollutants damaging the vegetation nearby and has serious environmental health impacts. The suspended particulate matter (SPM), SOx, soot, arsenic particles, cadmium, lead, etc. shoot up in the atmosphere near the smelters and the public suffers from several health problems.

(vi) *Occupational Health Hazards:* Most of the miners suffer from various respiratory and skin diseases due to constant exposure to the suspended particulate matter and toxic substances. Miners working in different types of mines suffer from asbestosis, silicosis, black lung disease, etc.

Remedial Measures: Safety of mine workers is usually not a priority subject of industry. Statistical data show that, on an average, there are 30 non-fatal but disabling accidents per ton of mineral produced and one death per 2.5 tons of mineral produced.

In order to minimise the adverse impacts of mining it is desirable to adopt eco-friendly mining technology. The low-grade ores can be better utilised by using microbial-leaching technique. The bacterium *Thiobacillus ferroxidans* has been successfully and economically used for extracting gold embedded in iron sulphide ore. The ores are inoculated with the desired strains of bacteria, which remove the impurities (like sulphur) and leave the pure mineral. This biological method is helpful from economic as well as environmental point of view.

Restoration of mined areas by re-vegetating them with appropriate plant species, stabilisation of the mined lands, gradual restoration of flora, prevention of toxic drainage discharge and conforming to the standards of air emissions are essential for minimizing environmental impacts of mining.

Green Resources

Forests are self-sustained wooden tracts with biotic community dominated by trees forming a sort by canopy. India gave birth to Gautam Buddha, in a sacred grove of "SAL Tress" dedicated to the Goddess "LUMBINI." He achieved salvation under a pipal tree forest in a peculiar organism of unlimited kindness and benevolences, that makes no demands for its sustenance and extend generosity through its products of life activity.

Areas Covered

India has a very rich Flora and Fauna much of which is present in forest area. The flora and Fauna of a forest are inter dependent on each other. The plant wealth of India's forests are made up of about 45000 species of trees, shrubs, herbs, creepers and climbers which account for about 12% of global plant wealth. Indian forest contain 7500 species of animals and as many as 60,000 species of insect pests.

The forests are complete ecosystem consisting of two components abiotic and biotic, i.e. both non-living and living components.

The Framework

The non-living factors prevailing in an ecosystem from abiotic components.

1. *Water:* It is a complex mixture of many compounds and gases. Mainly it is formed of two molecules hydrogen and oxygen. It forms 71% of the earth's surface and represents the most extensive habitat for the organisms. It helps in growth of the plants.
2. *Soil:* It is uppermost crust of the earth which is rich in organic and inorganic matter. It sustains the life of the plants on the earth. Plants absorbs essential nutrients from the soil-
3. *Climate:* It plays important role as it influences the dense growth, high yield and stimulates the other micro organism to form the hymns. Suitable climate of the region effects the forest canopy and increases forest products. It includes.
    (i) *Light:* It provides solar energy to the ecosystem for heating and photosynthesis. Besides photosynthesis, light controls morphogenesis. Photoperiods influence leaf fall, appearance of new leaves and flowering in plants.

    (ii) *Temperature:* Suitable temperature stimulates the rich growth of plants and transportation rate in which the plants give out excess water into atmosphere in the form of vapours. Most organisms survive a temperature range of 0° to 60° C less.

(iii) *Precipitation:* It is the process of condensation of water vapour which evaporate in the atmosphere through transpiration of forest trees. It helps in the formation of clouds, snow, hails, etc.

(iv) *Humidity:* Transpiration is inversely related to humidity. Humid areas have abundant dense plant, growth with numerous animals and epiphytes. Humidity maintains the soil moisture which helps in improving the quality of forests.

(v) *Wind:* It controls weather, transpiration, pollination and dissemination of propagules.

(vi) *Atmospheric Gases:* These include oxygen, nitrogen and carbondioxide. The carbonoxide is used in the process of photosynthesis. Nitrogen increases the fertility of the soil and oxygen is used for respiration thus, play important role in maintaining forest ecosystem.

The living organisms present in the forest form the biotic components. These include plants and animals. These are respectively called producers, consumers and decomposers.

(i) *Producers:* They are the plants in the form of grasses, herbs, shrubs and trees. These synthesise their own food by the process of photosynthesis. The inorganic material required are obtained from the forest ecosystem.

(ii) *Consumers:* These are mainly the animal which are unable to synthesise their food. Therefore, they take other organisms or parts of organisms. These are further divided into:

(a) *Herbivores (Primary Consumers):* These are the animals which feed on the green plants only e.g. deer, goat, rabbit.

(b) *Carnivores (Secondary Consumers):* These are animals which take herbivores e.g. frogs, snake, wolf.

(c) *Top Carnivores (Tertiary Consumers):* These are largest carnivores which feed on small carnivores e.g. Lion.

(iii) *Decomposers:* These are mainly bacteria and fungi, which obtain their food from organic materials of dead producers and consumers.

The Part

Forests are invaluable wealth of a country and a renewable natural resource. They are beneficial to man in several ways:

1. Forests are source of wood which is used as timber. Timber is the basic material for building houses making furniture, railways sleepers, sport goods, boats, ships, etc.

2. The forest provide fuel in the form of wood for cooking and keeping warm. Over 1500 million people use wood as a source of energy. Over 1000 million m^3 of wood is used as fuel in the world. In S-E. Asia fuel wood provides 42% of the total energy requirements. It is 58% of the total in Africa.

3. The forest provide food to the tribes in the form of tubers, roots, leaves and fruits.

4. Forest provides a large number of important products such as essential oils, (Kas sandal wood), natural camphor (Laurel tree-Lauru nobilis), tannins, resins, gums, dyes drug, spices, bidi wrappers (tendi leaves), soap substitutes like ritha and sikakai, lac, honey, wax, silk, etc. Horns hides, musk and ivory are provided by forest animals.

5. Bamboos form the poor's man timber. They are used in carts ropes, roofing, flooring, scaffolding, etc.

6. Forests keep the environment cool through increasing humidity of the atmosphere by transpiration which increases the chances of rainfall.
7. The fertility of the forest soil increases by dead fallen leaves which decay and form hymns.
8. Forests maintain oxygen contents of the atmosphere.
9. Forests reduce atmospheric pollution by using carbon dioxide in the process of photosynthesis.
10. Forest trees provide shelter to a number of animals such as insects reptiles, birds, mammal.
11. Forest prevent soil erosion by:
    (i) The roots which firmly binds the soil,
    (ii) Their fallen leaves which decrease the speed of water flow,
    (iii) Their trees which decrease the velocity of wind to prevent dislodging of the soil particle.
12. Forests have aesthetic importance as they simulate art of literature and music.
13. The humus of the forests floor acts as a giant sponge which absorbs large amount of water. This regulates the water flow in streams, springs, wells and rivers.

The Flora

According to Brewbaker (1984) total forest area of the world in 1900 was nearly 1000 million hectares (mha). By 1975 it was reduced to 2890 Mha and if this continues like this by 2000 A.D. it would be merely 2370 Mha. In India, nearby 80% of the country was covered with forests around 3000 B.C. Waves of migrants came to India, many from treeless countries and changed the whole landscape. There are no reliable statistics of forest cover in India. However according to Central Forestry Commission (1980), there are only 74.8 Mha of the land area under forest cover which forms about 22.7% of the total land mass of India. It has increased to 75.23 Mha (22.96%) in 1993.

In India there are 16 types of forests on the basis of geographical, climatic and edaphic conditions and their species composition. The most common in the tropical dry deciduous (38.7%) followed by tropical moist deciduous (30.9%) type. These two types of tropical deciduous forests contribute to the bulk (69.6) of the forest area in India.

As indicated earlier, according to CFC (1980) the forest cover in nearly 22.7%. However this is contrary to the general belief that effective forest cover is substantially low and that the forest wealth is deteriorating at rapid rate. Maximum damage in suffered by closed forests, which are more valuable from ecological and economic view points. We cannot afford depletion of these forests. The rapid decline of forests were:

(a) Overpopulation

(b) Deforestation

The Preservation

Deforestation is deterioration, reduction or removal of forest cover. It is the main cause of soil erosion in India. It is estimated that forest cover has declined from 7000 Mha in 1900 to 2890 Mha in 1985. It is expected that by 2000 A.D. forest cover will left only 2370 Mha. The reason for deforestation are given below.

Causes of Deforestation

1. *Population Explosion:* Population growth poses serious threat to the forest. The forests are the basis needs of everyday life as they provide us food, shelter and raw material for other essentialities but these forests deforested for fulfilling the increasing demands of over population like clearing of forests for agriculture, industries and urban area, etc.

2. *Fires:* Fire is worst enemy of the forests. Forests fire destroys adult as well as young trees and even seeds and useful fauna of the forests. Animal like is also lost and there is a danger to human life too in a forest fire.
3. *Pests:* Certain pests cause much damage in the forest. These either eat their leaves or bore into their stem or spread diseases.
4. *Industrial Development:* Industrial development although increases the economy of the country, but the basic raw material for these are obtained by the forests and ultimately growing industries for fulfilling their requirements started over exploiting the forests.
5. *Grazing and Growing Mammals:* These animals trample the seedlings or young plants, then destroy the leaves on the lower branches of tall trees and finally damage their trunks and roots.
6. *Agriculture Expansion:* Agriculture expansion means to convert the forest land area which are more fertile into agricultural areas to fulfil the crop necessities of the people of the country. Shifting cultivation also leads to deforestation in which a patch of forest land is cleared through lopping and burning the remainder. The ash is mixed in the soil. The cleared land is cultivated for one year and next year that land is left bared and another forest area cleared. This type of cultivation is called Jhum in north-eastern region, Podu in Andhra Pradesh bewar or dahza in M.P. Every year about 5 lakh hectares of land are cleared for shifting cultivation.
7. *Hydroelectric Projects:* Dams, reservoirs and hydroelectric projects submerge large forest tracts, killing all plants and animals of the area.
8. *Quarrying and Mining:* In forest and hilly areas mining and quarrying are extremely harmful as they spoil vegetation over large areas.

9. *Short Term Policies:* Short term policies made by our Government, i.e. five year plans also adversely affect the forest areas. It is easy to cut the forest areas in five years but not easy to grow in such a short tenure. Thus these policies play significant role in deforestation.

Effects of Deforestation

(i) Increased chances of soil erosion.

(ii) Floods and draughts become more frequent.

(iii) Man deprived of benefits of trees and woodland animals.

(iv) Increased chances of reduced rainfall but more dust storms.

(v) Extinction of forest dwelling species.

(vi) Economic life of forest dwelling people is adversely affected.

(vii) Pattern of rain fall is changing.

(viii) Scarcity of timber, fuel wood, etc. Shortage of fuel wood forces people burn animal dung which otherwise have been used as manure. This reduces soil fertility.

Greenery Secured

It is the development and management of forests in such a way to improve climate and water flow, prevent soil erosion and to provide optimum sustainable yield both for the present and future generations. The following steps can help conserve the forest wealth.

(i) Economically utilisation of timber to prevent wastage.

(ii) The wood fuel should be supplemented with alternative source of energy such as biogas.

(iii) Afforestation and reforestation should be undertaken. It is the development of forest where they did not exist in past (afforestation) or where they had been destroyed previously (reforestation). Afforestation should be done in areas unfit for agriculture, along highways and rivers, around playgrounds and parks and in spacious grounds. A special function of tree plantation, Van Mahotsava, is held each year in our country. In afforestation, native species should be preferred. Exotic species should be grown only on selective basis. NAEB (National Afforestation and Eco-development Board) looks after afforestation, tree plantation, restoration and development of ecologically fragile area.

(iv) Forest fires should be controlled through modern fire-fighting equipment.

(v) Using chemical and biological methods pests and diseases of forest trees should be controlled.

(vi) Grazing should be regulated and discouraged.

(vii) States forest policy should be adopted within overall frame work of National Forest Policy of 1988 which states that one third of total geographical area of country should be under forest cover.

(viii) Financial assistance to forestry sector should be increased as it was only 1.03% of total plan out lay in 7th five year plan and was further reduced to 0.94% in 8th five year plan.

Trees under Safety

It involves the improvement of our forests to ensure regular supply of useful plant products. There is requirement for improving many forest plants genetically for better yield, quick growth, resistance to pests and pathogens. It involves the following measures.

(i) *Improvement Cutting:* It involves cutting of all old dying trees, trees damaged by storms and animals, diseased trees and crooked trees.

(ii) *Selective Cutting:* It is the cutting of timber trees when they are fully matured. It is better to remove the less vigorous trees for timber and allow the others a chance for normal growth.

(iii) Forest floor should be quickly cleared off from fallen tree and broken branches.

(iv) Overcrowding of trees in the forest should be avoided.

(v) Afforestation Programmes.

Many afforestation programmes are undertaken by our government which are of three types.

Social Forestry Programme (1976): It was started in India in 1976 by NCA (National Commission on Agriculture). It involves the afforestation on public and common land for fuel wood, fodder, agricultural implements and fruits for the benefit of rural community. The programme is aimed at reducing the demand on existing forests. The plants selected for social forestry vary from place to place. The important ones are Albizia, Moringa, Delbergia, Acacia, Casuarina, Tectona, Zizyphus, Morus, etc.

Agro-forestry Programme: It is growing of multipurpose trees, shrubs, horticultural plants and forage plants alongwith crops. It is aimed at reviving the ancient practice of using the same land for fanning, forestry and animal husbandry.

Urban Forestry Programme: It is aimed at growing of fruit, flower and shade bearing shrubs and trees in urban areas along roads, in private compounds and in common parks. Forest conservation Act, 1980 was enforced to check indiscriminate deforestation.

Big Initiative

Just as man has important organ for breathing same is the case with earth. To save important part of the mother earth some people became conscious and started opposing the selfish people who not only destroyed environment of the earth but also destroyed themselves as well as other organisms. Chipko Andolan is similar movement in which people opposed the cruelties of man on the forest or nature. This Chipko Andolan started in the year 1972 in the upper reaches of Uttar Pradesh Himalayan region, Tehri Garhwal. Alaknanda valley faced unprecedented flood due to degradation of forests on the hills by the contractors. The menfolk opposed this large scale deforestation but never got success. The real movement started in March 1973 in Gopeshwar in Chamoli district when a sports good factory of Allahabad was to cut ten Ash trees near the village Mandal. The local people prevented the same by hugging the marked trees (hence chipko).

The same was repeated later in Rampur Phata in 1974, women led by Gaura Devi successfully prevented felling of forest trees near village Reni. The movement spread slowly to all nearly areas and outside by Chandi Prasad Bhatt of Gopeshwar and Sunder Lal Bahuguna of Silyara in Tehri region. Sunder Lal Bahuguna who is now a world famous figure on eco-protection conducted a peace march of about 3000 km in this hilly region to protest against the felling of trees and to prevent the construction of hydroelectric project in the silent valley region in Kerala. Such movements will help to preserve forests for future India. The Chipko movement succeeded and following are the results of this movement.

(i) Contractor system abolished by the Government.

(ii) Uttar Pradesh Forest Development Corporation department formed which works for the welfare of the hilly areas and the population residing there.

(iii) It slows down the process of deforestation.

(iv) It created awareness amongst the people of not only U.P. but of other states.

(v) It enlightened people about the necessity of ecological balance in nature.

8

Human Factors

Growth of Population

In 1800, the earth was home to about 1 billion people. The dramatic way in which global human population grew astonishing. It took about thirty-nine thousand years of human history to reach 1 billion, 130 years to reach the second billion, 45 years to reach 4 billion and the next doubling is likely within a span of a few decades. We have already crossed 6 billion and may reach 11 billion by 2045 as per the World Bank estimates.

Let us look at the reasons of this trend of human population growth. In the beginning of human civilisation, during the Stone Age, population was quite stable. Environmental conditions were hostile and humans had not yet developed adequate artificial means for adaptations to these stresses. Droughts and outbreak of diseases used to be quite common leading to mass deaths. The 14th century A.D. experienced large-scale mortality due to bubonic plague when about 50% of people in Asia and Europe died due to the disease.

With scientific and technological advancement, life expectancy of humans improved. People started living in definite settlements leading a more stable life with better sanitation, food and medical facilities. Victory over famine-related deaths and infant mortality became instrumental for a rapid increase in population size. In agriculture-based societies children were considered as economic assets who would help the parents in the fields and that is why in the developing countries, population growth climbed to unthought-of heights, at the rate of 3-4% per year, accounting for about 90-95% of total population growth of the world in the last 50 years.

Features of Population

Exponential Growth: When a quantity increases by a constant amount per unit time e.g. 1, 3, 5, 7, etc. it is called linear growth. But, when it increases by a fixed percentage it is known as exponential growth e.g. 10, 10^2, 10^3, 10^4, or 2, 4, 8, 16, 32, etc. Population growth takes place exponentially and that explains the dramatic increase in global population in the past 150 years.

Doubling Time: The time needed for a population to double its size at a constant annual rate is known as doubling time. It is calculated as follows:

Td = 70/r

where Td = Doubling time in years

r = annual growth rate

If a nation has 2% annual growth rate, its population will double in 35 years.

Total Fertility Rates (TFR): It is one of the key measures of a nation's population growth. TFR is defined as the average number of children that would be born to a woman in her lifetime if the age specific birth rates remain constant. The value of TFR varies from 1.9 in developed nations to 4.7 in developing nations. In 1950's the TFR has been 6.1. However, due to changes in cultural and technological setup of societies and government policies the TFR has come down which is a welcome change.

Infant Mortality Rate: It is an important parameter affecting future growth of a population. It is the percentage of infants died out of those born in a year. Although this rate has declined in the last 50 years, but the pattern differs widely in developed and developing countries.

Replacement Level: This is an important concept in population dynamics or demography. Two parents bearing two children will be replaced by their offspring. But, due to infant mortality this replacement level is usually changed. For developing nations, where infant mortality is high and life expectancy is low, the replacement level is approx 2.7, whereas in developed nations it is 2.1.

Age Structure: Age structure of population of a nation can be represented by age pyramids, based upon people belonging to different age classes like pre-reproductive (0-14 years), reproductive (15-44 years) and post-reproductive (45 years and above). We get three types of age pyramids:

(a) *Pyramid Shaped:* Here the very young population is more, making a broad base and old people are less. This type indicates growing population. India, Bangladesh, Ethiopia, Nigeria are examples of this type. The large number of individuals in very young age will soon enter into reproductive age, thus causing an increase in population, whereas less number of people in old age indicate less loss of population due to death.

(b) *Bell Shaped:* It occurs in countries like France, USA and Canada where birth rates have in the past one or two decades declined resulting in people of almost equal number in age group 0-35 years. So in the next 10 years, the people entering into reproductive age group is not going to change much and such age-pyramids indicate stable populations.

(c) *Urn Shaped:* Here number of individuals in very young class is smaller than the middle reproductive age class. In the next 10 years the number in reproductive age class will thus become less than before resulting in a decline of population growth. Germany, Italy, Hungary, Sweden and Japan are examples of this type.

The TFR, age structure, infant mortality and replacement level are all important parameters determining population growth. But population will not stop growing even when all couples have only 2 children.

Case Studies: Ethiopia is a developing nation with a pyramid-shaped age structure indicating expanding population. Its TFR is 6.9 presently. Even if it aims to reach the replacement level by the year 2050, its population that is 57 million now would rise to 225 million by 2050 when TFR becomes 2.1 and continue growing until it levels off, 100 years later, at 370 million. Population growth is also affected due to AIDS in the HIV-prevalent countries mainly in Africa.

The earlier population projections of UN are now found to be reduced by 8% in the seriously HIV-affected countries, i.e. Mali, Rwanda, Uganda and Zambia. In Zimbabwe, HIV affects a quarter of the population aged between 15-49 yrs. In Botswana, 2/3rd of the 15-year-olds are predicted to die of AIDS before reaching 50 years of age. About 30% of adult population in many African countries is HIV-positive. This has drastically reduced life expectancy in these countries.

Zero Population Growth (ZPG): When birth plus immigration in a population are just equal to deaths plus emigration, it is said to be zero population growth.

Male-female Ratio: The ratio of boys and girls should be fairly balanced in a society to flourish. However, due to female infanticides and gender-based abortions, the ratio has been upset in many countries including India. In China, the ratio of boys to girls became 140:100 in many regions which led to scarcity of brides.

Life Expectancy: It is the average age that a newborn infant is expected to attain in a given country. The average life expectancy, over the globe, has risen from 40 to 65.5 years over the past century. In India, life expectancy of males and females was only 22.6 years and 23.3 years respectively in 1900. In the last 100 years improved medical facilities and technological advancement has increased the life expectancy to 60.3 years and 60.5 years, respectively for the Indian males and females. In Japan and Sweden, life expectancy is quite higher, being 82.1-84.2 for females and 77-77.4 for males, respectively.

Demographic Transition: Population growth is usually related to economic development. There occurs a typical fall in death rates and birth rates due to improved living conditions leading to low population growth, a phenomenon called demographic transition.

It is associated with urbanisation and growth and occurs in four phases:

(a) Pre-industrial phase characterised by high growth and death rates and net population growth is low.

(b) Transitional phase that occurs with the advent of industrialisation providing better hygiene and medical facilities and adequate food, thereby reducing deaths. Birth rates, however, remain high and the population shows 2.5-3% growth rate.

(c) Industrial phase while there is a fall in birth rates thereby lowering growth rate.

(d) Post industrial phase during which zero population growth is achieved.

Demographic transition is already observed in most developing nations. As a result of demographic transition the developed nations are now growing at a rate of about 0.5% with a doubling time of 118 years. However, the matter of concern is that more than 90% of the global population is concentrated in developing nations which have a growth rate a little more than 2%, and a doubling time of less than 35 years.

Population Explosion: There has been a dramatic reduction in the doubling time of the global human population, as we have already discussed. In the 20th century, human population has grown much faster than ever before. Between 1950-1990, in just 40 years the population crossed 5 billion mark with current addition of about 92 million every year, or so to say, adding a new Mexico every year. In the year 2000, the world population was 6.3 billion and it is predicted to grow four times in the next 100 years. This unprecedented growth of human population at an alarming rate is referred to as population explosion.

The Indian Scenario: India is the second-most populous country of the world with 1 billion people. If the current growth rates continue, it will have 1.63 billion people by 2050 and will become the most populous country surpassing China. So we are heading for very serious ramifications of the population explosion problem. Do we have the resources and provisions for feeding, housing, educating and employing all those people being added every year? If we look at the population statistics of our country we find that in just 35 years after independence we added another India in terms of population. On 11th May, 2000 we became 1 billion and now we can say that every 6th person in this world is an Indian.

The Population Clock: Every second, on an average 4-5 children are born and 2 people die, thus resulting in net gain of nearly 2.5 person every second. This means that every hour we are growing by about 9000 and everyday by about 2,14,000.

Population explosion is causing severe resource depletion and environmental degradation. Our resources like land, water, fossil fuels, minerals, etc. are limited and due to over-exploitation these resources are getting exhausted. Even many of the renewable resources like forests, grasslands, etc. are under tremendous pressure. Industrial and economic growth are raising our quality of life but adding toxic pollutants into the air, water and soil.

As a result, the ecological life-support systems are getting jeopardised. There is a fierce debate on this issue as to whether we should immediately reduce fertility rates through world-wide birth control programs in order to stabilise or even shrink the population or whether human beings will devise new technologies for alternate resources, so that the problem of crossing the carrying capacity of the earth will never actually come. There are two very important views on population growth which need a mention here:

Malthusian Theory: According to Malthus, human populations tend to grow at an exponential or compound rate whereas food production increases very slowly or remains stable. Therefore, starvation, poverty, disease, crime and misery are invariably associated with population explosion. He believes "positive checks" like famines, disease outbreak and violence as well as "preventive checks" like birth control need to stabilise population growth.

Marxian Theory: According to Karl Marx, population growth is a symptom rather than the cause of poverty, resource depletion, pollution and other social ills. He believed that social exploitation and oppression of the less-privileged people leads to poverty, overcrowding, unemployment, environmental degradation that in turn, causes over-population.

A compromise between the two views is required because all these factors seem to be interdependent and interrelated. Equity and social justice to all, allowing everyone to enjoy a good standard of living is the need of the hour that can voluntarily help in achieving a stabilised global population.

Family Welfare Programmes: Population explosion is like a time bomb that must be diffused well in time. The population must be kept much below the carrying capacity and stabilised, so that the aftermath of explosion could be avoided.

It is not precisely known as to how long can we continue our exponential growth in population and resource use without suffering overshoot or dieback. We are getting warning signals that if not controlled, the increasing population is going to deplete all the resources beyond their regeneration capacity. A catastrophic doomsday model warns us that the earth cannot sustain more than two more doublings, *i.e.* 25 billion.

The United Nations projections about population stabilisation of developed and developing nations and that of Asia are alarming. The ratio is derived by dividing crude birth rate by crude death rate. As evident, developed nations have already achieved a stabilisation ratio of 1 around the year 2000, which is more or less stabilised indicating zero population growth. Developing nations including Asia, on the other hand, is yet having a high stabilisation ratio nearing 3, which is however, on a decline and is expected to lower down substantially by 2025. Stabilisation in developing nations is possible only through various family welfare programmes.

Adopting Kerala Model (A case study): Kerala has earned the distinction of having lowest birth rates among all the states of India. The main parameters deciding the effectivity of this model depends upon the age of marriage for women at 21 years, as against and Indian average of 18 years, female literacy of 53% against Indian average of 13%, greater emphasis on primary education with 60% budget provision for the same, as against 50% in many other states, better public distribution system of food among 97% of population, better medical facilities in rural areas and greater success of family planning programmes. The Kerala Model has its own success story emphasizing the effectivity of social justice approach for family planning.

Family Planning: Family planning allows couples to decide their family size and also the time spacing of their offspring. Almost every culture in the past used to practise some traditional fertility control methods through some traditions, taboos and folk medicine.

Modern science has provided several birth control techniques including mechanical barriers, surgical methods, chemical pills and physical barriers to implantation. More than a hundred contraceptive methods are on trial. The United Nations Family Planning Agency provides funds to 135 countries. Many of these countries include abortion as a part of the population control programme which very often encourages female infanticide thereby disturbing the optimal male: female ratio in a society. The birth control programmes have often faced strong opposition from religious groups.

Nonetheless, World Health Organisation (WHO) estimates that today about 50 per cent of the worlds' married couples adopt some family planning measures as compared to just 10% about 30 years back. Still some 300 million couples do not have access to family planning.

The Indian Scenario: India started the family planning programme in 1952 while its population was nearly 400 million. In 1970's, forced family planning campaign by the Government resulted in a turmoil all over the country.

In 1978, the government raised the legal minimum age of marriage from 18 to 21 for men and 15 to 18 years for women. Even in 1981 census no drop in population growth was observed. Since then funding for family planning programmes has been increased further.

Unable to reach a consensus regarding population policy, the state governments in 2000 were allowed to adopt their own approach. In Kerala, the population has been stabilised with a focus on social justice as already discussed. It is now comparable to many industrialised nations including USA and it has proved that wealth is not a prerequisite for zero population growth. Andhra Pradesh has also just achieved the target of ZPG in 2001, but it has been done with a different approach. The poor class was encouraged to be sterilised after two children by paying cash incentives, better land, housing, wells and subsidised loans. In contrast, Bihar and UP have showed increase in their growth rates (more than 2.5%).

Successful family planning programs need significant societal changes including social, educational and economic status for women, social security, political stability, proper awareness and confidence building along with accessibility and effectivity of the birth control measures.

Environment and Human Health: In general terms a physically fit person not suffering from any disease is called a healthy person. However, there are many other dimensions associated with the state of being healthy. According to World Health Organisation (WHO) health is "a state of complete physical, mental and social well-being and not merely the absence of disease or infirmity". Human health is influenced by many factors like nutritional, biological, chemical or psychological. These factors may cause harmful changes in the body's conditions called disease.

Infectious Organisms: Disease-causing organisms pose greater environmental threats to health, more severely in the developing countries especially the tropical ones. High temperature and moisture along with malnutrition help many diseases to spread in these countries. Microbes especially bacteria can cause food poisoning by producing toxins in the contaminated food. Some moulds grow on food and produce poisonous toxins.

Infectious organisms can also cause respiratory diseases (pneumonia, tuberculosis, influenza, etc.) and gastrointestinal diseases (diarrhoea, dysentery, cholera, etc.).

There are various types of parasites that cause diseases like malaria, schistosomiasis, filariasis, etc. Most of these infections take place when the environmental conditions are unclean and unhygienic.

Chemicals: A large number of chemicals are introduced in the environment by anthropogenic activities. Industrial effluents containing various chemicals are of major concern. Chemicals can be divided into two categories, i.e. hazardous and toxic chemicals. Hazardous are the dangerous chemicals like explosives, inflammable chemicals, etc. Toxic chemicals (toxins) are poisonous chemicals which kill cells and can cause death. Many other chemicals can cause cancer (carcinogenic), affect genetic material (DNA) in cells (mutagenic) or cause abnormalities during embryonic growth and development (teratogenic), while there are others that affect nervous system (neurotoxins) and the reproductive system.

Some of the pesticides and other industrial pollutants may act as hormone analogs in humans and other species. These environmental hormones affect reproduction, development and cause various types of ailments including tumors.

Many chemicals like DDT and other chlorinated pesticides bio-accumulate in food-chain and show deleterious effects at the top of the food chain. Many chemical substances present in waste waters like heavy metals (mercury, cadmium, lead, etc.) fluoride and nitrate can affect human health. The adverse effects of some of these have already been discussed in Unit 5. Metals can contaminate food while cooking in various types of utensils including alloys like steel.

Containers for canned food, especially which are acidic in nature, contaminate the food with lead. Lead also comes in water from the water-pipes where it is added for plumbing purposes. Various alcoholic beverages contain lead while tobacco contains cadmium that goes in the body and affects human health.

Various chemicals, gases and particulates laden with chemicals, spewed into the environment from various industries cause air pollution and affect human health. The details of effects of air pollution on human health have already been given in Unit 5.

Noise: Although human ear is capable of tolerating a range of sound levels, yet if sound levels beyond the permissible level exist for certain duration, it becomes painful and sometimes irreparable damage occurs. Besides hearing damage various types of physiological and psychological changes are induced by noise pollution. The details of effects of noise on human health are given in Unit 5.

Radiations: Radiations are known to cause short-term and long-term changes in various organs. Cosmic rays and ultra-violet rays cause harmful effects on human health which may include cancer. The details of ill effects of radiations on human health are given in Unit 5.

Diet: Diet has a very important role in maintaining health. Malnutrition makes humans prone to other diseases. There is a strong correlation between cardiovascular diseases and the amount of salt and fat in one's diet. Food contamination can cause various ill effects.

There had been cases of Dropsy in India, a disease which occurred due to contamination of mustard oil with the poisonous seeds of *Argemone mexicana.* Likewise various adulterated pulses, condiments, oils, etc. sold in the market to earn profit affect human health.

Settlement: Proper environment, availability of basic necessities of life like, water, sanitation, etc. are essential for healthy living. Housing is very important from security point of view. Improper settlement and poor physical environment may cause various psychological problems which affect various vital physiological processes in the body.

Concern of Human Rights

Human rights are the rights that a human being must enjoy on this earth since he/she is a human being. Although the foundation of human rights was laid in the 13th century when resistance to religious intolerance, socio-economic restraints and scientific dogmas resulted in some revolts mainly due to the liberal thoughts of some philosophers. However, true hopes for all people for happy, dignified and secure living conditions were raised with the Universal Declaration of Human Rights (UNDHR) by the UNO on December 10, 1948. This declaration provided comprehensive protection to all individuals against all forms of injustice and human rights violations.

The UNDHR defines specific rights, civil, political, economic, social as well as cultural. It defines the rights to life, liberty, security, fair trial by law, freedom of thought, expression, conscience, association and freedom of movement. It emphasises right to equal pay for equal work, right to form and join trade unions, right to health care, education, adequate rest, etc.

Although the human rights are considered to be universal, there is a wide disparity between the developing and the developed countries. Population and poverty are often found to be the most important causes of violation of human rights in the third world countries. Poverty often undermines human dignity and without dignity there is no meaning of human right. In fact, talks of human rights seem justified only when one can just manage to live on.

The World Health Organisation estimates indicate that one out of every five persons in this world is malnourished, lacks clean drinking water, lacks proper hygienic conditions and adequate health facilities; one out of three persons does not have enough fuel to cook or keep warm and one out of five persons is desperately poor for whom life is nothing but struggle for survival. Every year 40 million people are dying due to consumption of contaminated drinking water. There is acute scarcity of employment in the Third World countries. Under such conditions, a poor man feels that perhaps his child can earn something for himself or the family. For him, the merit of universal education and child labour prevention is of much less importance than his grim struggle for existence.

For the developed countries, which have already attained a high stage of development in material and economic resources, the social and economic rights are not that important as civil and political rights. Whereas, the reverse is true for the developing countries which are struggling for life under conditions of extreme poverty, ignorance, illiteracy, malnutrition and diseases. For them the civil and political rights carry little meaning.

In June 1993, during the Vienna World Conference on Human Rights the need for economic and social rights were considered as equal to the West's political and civil rights. Respect towards human rights is now considered to be one of the important criteria for giving development assistance to a country. In 1992, the Burton Bill passed in USA slashed 24 million dollars of development assistance to some developing countries including India on the grounds of showing poor human rights records.

In India, human right issues have mostly centered around slavery, bonded labour, women subordination, custodial deaths, violence against women and minorities, child abuse, dowry deaths, mass killings of dalits, torture, arbitrary detentions, etc. The Constitution of India contains a long list of people's civil, political, economic and social rights for improving their life. Yet, it is an irony that violation of human rights takes place rather too often in our country.

Social discriminations, untouchability, patriarchal society with male domination, etc. still prevail in the society which hinder the honour of human rights. Civil liberties and fundamental freedom are also often violated by those who have money power. Communal violence against minorities has become quite prevalent in our country. There is a need to respect the human rights of all people in every nation for overall development and peace.

As the right to development was defined, another aspect of human right related to environment emerged. After the Earth Summit 1992, the need for sustainable development was recognised. Soon after on May 16, 1994 at Geneva, the United Nations drafted the first ever Declaration of Human Rights and Environment, which embodies the right of every human being to a healthy, secure and ecologically sound environment. A sustainable society affirms, equity, security, attainment of basic human needs and environmental justice to all.

It is quite disheartening to look at the environmental inequities. The developed nations utilizing most of the natural resources and reaping the benefits of industrial development are not bearing the burden of their hazardous wastes, as they export such wastes to many developing countries who have to face the toxic impacts of the hazardous wastes.

The worker class and the poor are the main victims and sufferers of adverse effects of industrial toxins, foul smelling polluted air, unclean and unsafe drinking water, unhealthy working conditions, occupational health hazards, etc. The indigenous people and tribal people are the worst victims of development who lose their homes and lands to dams and reservoirs and are deprived of their human rights to native homes.

Draft Declaration of Human Rights and Environment: The draft declaration describes the rights as well as duties that apply to individuals, governments, international organisations and transnational corporations.

The preamble envisages a deep concern regarding the consequences of environmental harm caused by poverty, debt programmes and international trade. Environmental damages are often irreversible. Human rights violations may lead to further environmental degradation on a long-term basis and the environmental degradation in turn would lead to further human rights violation. The principles of the draft declaration are divided into five parts.

Part I: It deals with human rights for an ecologically sound environment, sustainable development and peace for all. It also emphasises the present generation's rights to fulfill its needs to lead a dignified and good quality life. But, at the same time it lays stress on the fact that it should be without impairing the rights of the future generations to meet their needs.

Part II: It mainly deals with human rights related to an environment free from pollution and degradation. It also emphasises the rights to enjoyment of natural ecosystems with their rich biodiversity. It defines right to own native land or home. No one can be evicted from one's native place except in emergency or due to a compelling purpose benefitting the society as a whole which is not attainable by other means. All persons have the right to timely assistance in the event of any natural or technological disaster.

Part III: It deals with right of every person to environmental information, education, awareness and also public participation in environmental decision making.

Part IV: It deals with the duties to protect and preserve the environment and prevent environmental harm. It includes all remedies for environmental degradation and measures to be taken for sustainable resource use. It emphasises that states shall avoid using environment as a means of war and shall respect international law for protection of environment.

Part V: This lays stress on social justice and equity with respect to use of natural resources and sustainable development.

Till now, however, it has not been defined in practical terms the threshold, below which level of environmental quality must fall before a breach of individual human right is said to have occurred or above which the level of environmental quality must rise. 'Right to development' has to be linked to 'right to safe and clean environment' which has to be considered not only at the level of individual but at community, national and global level.

Concerns of Value Education: Education is one of the most important tools in bringing about socio-economic and cultural progress of a country. However, the objective of education should not merely be imparting coaching to the students that they get through the examinations with good results and get some good job. Education does not simply mean acquiring a lot of information but also its righteousness and use within the framework of a spectrum of ethical values.

The rapid strides of scientific and technological advancements have no doubt, brought revolutionary changes in our every day life and information technology has shrunk the whole world into a "global village", with access to very information sitting in one corner over the internet. But, in this frenzy for development and mad race for progress perhaps man has become too materialistic, self-centered and overambitious and the desired ideals of a real good life have been pushed to the background.

Value-based education thus has a very significant role in providing proper direction to our youth, to inculcate a positive attitude in them and to teach them the distinction between right and wrong. It teaches them to be compassionate, helpful, peace loving, generous and tolerant so that they can move towards a more harmonious, peaceful, enjoyable and sustainable future.

Value education helps in arriving at value-based judgements in life-based on practical understanding of various natural principles rather than acquiring certain prejudices. Value education encompasses human values, social values, professional values, religious values, national values, aesthetic values and environmental values. Value education increases awareness about our national history, our cultural heritage, national pride, constitutional rights and duties, national integration, community development and environment.

Value education has different phases, *i.e.* value awareness, value orientation, value appraisal, value selection, value commitment and value action. The basic aim is to create and develop awareness about the values, their significance and role. After knowing them the student's mindset would get oriented towards those values and he will try to critically analyse the same and then select the values which really appeal to him. This will be followed by commitment that needs to be reaffirmed over and over again so that every action is taken keeping those values in view.

Environmental Education on the basis of Value: Environmental education or environmental literacy is something that every person should be well versed with. The principles of ecology and fundamentals of environment can really help create a sense of earth-citizenship and a sense of duty to care for the earth and its resources and to manage them in a sustainable way so that our children and grand children too inherit a safe and clean planet to live on.

We have already discussed environmental ethics, earth-citizenship and ways and means to propagate environmental education and awareness. Following the Supreme Court directives (in M.C. Mehta Vs. Union of India, 1988) environmental education has been included in the curriculum right from the school stage to college/ university level. The prime objective of the same is to make everyone environment literate. The environment belongs to each one of us and our actions affect the environment. When the environment gets degraded it affects our health, well-being and our future. So we have a right to know the a b c of environment and also have a right to safe and clean environment.

Let us now see how environmental education be made value-based:

1. Preparation of text books and resource materials about environmental education can play an important role in building positive attitudes about environment. The basic human value 'man in nature' rather than 'nature for man' needs to be infused through the same.
2. Social values like love, compassion, tolerance and justice which are the basic teachings of most of our religions need to be woven into environmental education. These are the values to be nurtured so that all forms of life and the biodiversity on this earth are protected.
3. Cultural and religious values enshrined in Vedas like *"Dehi me dadami te"*, i.e. "you give me and I give you" (Yajurveda) emphasise that man should not exploit nature without nurturing her. Our cultural customs and rituals in many ways teach us to perform such functions as would protect and nurture nature and respect every aspect of nature, treating them as sacred, be it rivers, earth, mountains or forests.
4. Environmental education should encompass the ethical values of earth-centric rather than human-centric world-view. The educational system should promote the earth-citizenship thinking. Instead of considering human being as supreme we have to think of the welfare of the earth.
5. Global values stress upon the concept that the human civilisation is a part of the planet as a whole and similarly nature and various natural phenomena over the earth are interconnected and inter-linked with

special bonds of harmony. If we disturb this harmony anywhere there will be an ecological imbalance leading to catastrophic results.

6. Spiritual values highlight the principles of self-restraint, self-discipline, contentment, reduction of wants, freedom from greed and austerity. All these values promote conservationism and transform our consumeristic approach.

The above-mentioned human values, socio-cultural, ethical, spiritual and global values incorporated into environmental education can go a long way in attaining the goals of sustainable development and environmental conservation. Value-based environmental education can bring in a total transformation of our mind set, our attitudes and our life-styles. "What is the use of building a beautiful house if you don't have a decent planet to place it on?"-perhaps this single question can answer the main burning question-"What is real development and progress?" We certainly do not want development in exchange of environmental disasters, health hazards, loss of mental peace and merciless destruction of nature's beauty and natural resources. The value elements in environmental education alone can succeed in achieving the real goals of environmental literacy.

HIV/AIDS: AIDS, the Acquired Immuno Deficiency Syndrome is not a hereditary disease but is caused by HIV (Human Immunodeficiency Virus). HIV from an infected person can pass to a normal person through blood contact generally during unprotected sex with infected person and sharing needles or syringes contaminated with small quantities of blood from HIV positive person. HIV can also pass from infected mothers to their babies during pregnancy, delivery or breast-feeding. HIV, however, doesn't spread through tears, sweat, urine, faeces or saliva during normal kissing. It also does not spread by sharing utensils, towels, clothing, toilet seats or insect bite like that of mosquito or bed bug.

According to a recent estimate about 40 million people are living with HIV/AIDS worldwide and 70% of them in Sub Saharan Africa. HIV/AIDS has been identified as the fourth largest cause of mortality. About 3 million people died due to HIV/AIDS in 2003. AIDS is rapidly spreading in eastern Europe and Asia. It is expected that in the coming decades there will be sharp increase in HIV/AIDS cases in Russia, China, and India.

AIDS was discovered in 1983. Although sufficient knowledge has been gained about the disease, a definite source of this virus could not be identified.

Most evidences have suggested that AIDS has spread from Africa. It is believed that the virus has been transferred to humans from primates like African Monkey (White sooty mangabeys) or chimpanzees.

According to another theory HIV has spread through vaccine programmes in various parts of the world in the following manner:

1. HIV has spread in Africa through HIV contaminated polio vaccine prepared by using monkey's kidney.
2. It had spread through hepatitis B viral vaccine in New York, Los Angeles and San Francisco.
3. It has spread through small pox vaccine programme of Africa. It is also hypothesised that AIDS is a man-made epidemic produced by genetically engineered laboratory produced virus. AIDS itself does not kill humans. The deaths occur due to attack by other diseases because of the weakening of immune system. There is decline in T-cells which are the key infection fighters in the immune system. HIV destroys or disables these cells as a result of which various types of infectious diseases due to microbial invasion occur. Even dreaded disease like cancer can easily develop in the HIV infected persons.

Consumption of alcohol is understood to increase the susceptibility to infection and progression of AIDS.

Influence of HIV/AIDS on Environment: When there is an AIDS epidemic large number of deaths occur which adversely affect local environment and natural resources. Due to large number of deaths there is loss of labour and the level of production decreases. With fewer adults, young members with limited resources like land and lack of experience and knowledge find it difficult to look after the perennial crops and prefer crops requiring less labour and time.

They devote less time for soil conservation, forestry conservation, especially if there are deaths of professional forest workers. Demand of easily accessible fuel wood increases. More timber is required for making coffins or for pyre making. More water is required for maintaining hygiene in AIDS affected locality. The HIV carriers are also not able to perform well due to lack of energy and frequent fever and sweating.

Women and Children: Well-being: Women and children are usually the soft targets, who suffer in a number of ways mainly because they are weaker, helpless and economically dependent.

Welfare of Women: Women usually suffer gender discrimination and devaluation at home, at workplace, in matrimony, in inheritance, in public life and power, particularly in developing countries. The gender violence, victimisation and harassment take many forms across culture, race or nation. The statistical data provided by the Ministry of Women and Child Development is an eye-opener that deglorifies the celebrated culture of our country.

The exceptionally high number of cases of abduction, dowry deaths, rape, domestic violence, criminal offences and mental torture to women is something that needs immediate attention and reforms in the interest of the women. Women are often the worst victims of communal enemities. The human rights of women are violated too often in a male-dominated patriarchal society.

Thus, there is an urgent need for policy reforms and more stringent legislation as well as educational and legal awareness amongst women for checking the atrocities and injustice towards her. There are now many 'Women Groups' who actively take up women welfare issues and legally constituted "Women cells" that exist almost everywhere and fight for protection of women rights and dignity. There is a full-fledged Ministry for Women and Child development whose sole aim is to work for the welfare and upliftment of women encompassing family planning, health care, education and awareness. There is a need for complete transformation and re-orientation of social ethos for restoring the dignity, status, equality and respect for women.

Women are also the victims of capitalism, development and environment. The exploitative nature of capitalist development not only affects the natural environment but the traditional, social, cultural and family life of women. After losing the forests and getting de-habilitated from their native places, men folk usually migrate to towns in search of some job while the women are left behind to look after the family and household with little resources. Development projects like mining very often play havoc with the life of women. Men can still work in the mines or migrate to towns after getting compensation from the government.

The National Network for Women and Mining (NNWM) with about 20 groups in different mining states of India is rightly fighting for a "gender audit" of India's mining companies. The displaced women are the worst affected as they do not get any compensation and are totally dependent upon the males for wages. The displaced women driven out from their land-based work are forced to take up marginalised work which is highly unorganised and often socially humiliating. Issues related to their dignity and honour have not yet received any attention. The NNWM is now working for rights of women over natural resources, resettlement and compensation issues.

Besides the government initiatives there are now a number of nongovernment organisations (NGO's), mostly as "Mahila Mandals" to create awareness amongst women of remote villages even to empower them, train them, educate them and help them to become economically self-dependent.

On an international level, the United Nations Decade for Women (1975-85) witnessed inclusion of several women welfare related issues on international agenda. The CEDAW (International Convention on the Elimination of All forms of Discrimination Against Women, 1979) has been a landmark outcome of the decade to be accepted as an international standard for the protection and promotion of women's human rights and socio-economic upliftment. It is, however, most important for all women, in the mainstream, tribals, refugees and the downtrodden to be educated about these issues.

Welfare of Children: Children are considered to be the assets of a society. But, ironically, the statistical figures tell us that about a million babies, out of 21 million born every year in India are abandoned soon after their birth due to different socio-economic reasons.

Around 20 million children in our country are estimated to be working as child labours, some of them in various hazardous industries like the match industry, firework industry, brassware industry and pottery industry. Poverty is the main reason to drive these children into long hours of work in miserable, unhealthy conditions and yet they do not get the minimum nutritive food, what to talk of educational and recreational facilities, which are their childhood rights.

The UN General Assembly in 1959 adopted the Declaration of the Rights of a child. After the UN convention on Rights of Child, it became International Law in the year 1990, consisting of 54 articles and a set of international standards and measures to promote and protect the well-being of children in a society.

The law defines right of the child to survival, protection, development and participation. The right to survival emphasises on adequately good standards of living, good nutrition and health. The right to protection means freedom from exploitation, abuse, inhuman treatment and neglect. The right to development ensures access to education, early childhood care and support, social security and right to leisure and recreation. The right to participation means freedom of thought, conscience and religion and appropriate information to the child.

The World Summit on Children, held on September 30, 1990 had a focussed agenda for the well-being of the children targeted to be achieved in the beginning of the new millennium. India is also a signatory to the World Declaration on Survival, protection and development of children. A national plan of action for children has been formulated by the Ministry of Human Resource Development (MHRD), Government of India in which a strategic plan has been formulated for children's welfare in the priority areas of health, education, nutrition, clean and safe drinking water, sanitation and environment. Universalisation of effective access to at least primary level schooling, special emphasis on girl child's education including health and nutrition, upgradation of home-based skills, mid-day meals scheme, expansion of early childhood development activities including low-cost family based involvements are some of the important actions envisaged.

Children are also the most affected due to environmental pollution. "They consume more water, food and air than adults, hence more susceptible to any environmental contamination" says one of the scientific reports of Centre for Science and Environment (CSE), New Delhi. Water-borne diseases are the biggest threat to children, affecting around 6 million children in India. Childhood cancer rates are also increasing by 6% every year. Even the growing foetus in the mother's womb is not safe and free from the adverse effects of environmental toxins. It is high time to work together for a secure and cleaner environment so as to give our children a cleaner and safer world to live in.

Part of Information Technology in Environment and Human Health: Information technology has tremendous potential in the field of environmental education and health as in any other field like business, economics, politics or culture. Development of internet facilities, worldwide web, geographical information system (GIS) and information through satellites has generated a wealth of up-to-date information on various aspects of environment and health. A number of soft-wares have been developed for environment and health studies which are user-friendly and can help an early learner in knowing and understanding the subject.

Database: Database is the collection of inter-related data on various subjects. It is usually in computerised form and can be retrieved whenever required. In the computer the information of database is arranged in a systematic manner that is easily manageable and can be very quickly retrieved. The Ministry of Environment and Forests, Government of India has taken up the task of compiling a database on various biotic communities. The comprehensive database includes wildlife database, conservation database, forest cover database, etc. Database is also available for diseases like HIV/AIDS, Malaria, Fluorosis, etc. National Management Information System (NMIS) of the Department of Science and Technology has compiled a database on Research and Development Projects along with information about research scientists and personnel involved.

Environmental Information System (ENVIS): The Ministry of Environment and Forests, Government of India has created an Information System called Environmental Information System (ENVIS). With its headquarters in Delhi, it functions in 25 different centres all over the country. The ENVIS centers work for generating a network of database in areas like pollution control, clean technologies, remote sensing, coastal ecology, biodiversity, western ghats and eastern ghats, environmental management, media related to environment, renewable energy, desertification, mangroves, wildlife, Himalayan ecology, mining, etc. The National Institute of Occupational Health provides computerised information on occupational health, i.e. the health aspects of people working in various hazardous and non-hazardous industries, safety measures, etc.

Remote Sensing and Geographical Information System (GIS): Satellite imageries provide us actual information about various physical and biological resources and also to some extent about their state of degradation in a digital form through remote sensing. We are able to gather digital information on environmental aspects like water-logging, desertification, deforestation, urban sprawl, river and canal network, mineral and energy reserves and so on. Geographical Information System (GIS) has proved to be a very effective tool in environmental management. GIS is a technique of superimposing various thematic maps using digital data on a large number of inter-related or interdependent aspects. Several useful softwares have been developed for working in the field of GIS. Different thematic maps containing digital information on a number of aspects like water resources, industrial growth, human settlements, road network, soil type, forest land, crop land or grassland, etc. are superimposed in a layered form in computer using softwares.

Such information is very useful for future land-use planning. Even interpretations of polluted zones, degraded lands or diseased cropland, etc. can be made based on GIS.

Planning for locating suitable areas for industrial growth is now being done using GIS by preparing Zoning Atlas. GIS serves to check unplanned growth and related environmental problems. Our satellite data also helps in providing correct, reliable and verifiable information about forest cover, success of conservation efforts, etc.

They also provide information of atmospheric phenomena like approach of monsoon, ozone layer depletion, inversion phenomena, smog, etc. We are able to discover many new reserves of oil, minerals, etc. with the help of information generated by remote sensing satellites. Thus remote-sensing and GIS play a key role in resource mapping, environmental conservation, management, planning and environmental impact assessment.

It also helps in identifying several disease-infested areas which are prone to some vector-borne diseases like malaria, schistosomiasis, etc. based upon mapping of such areas.

There are several Distribution Information Centres (DICs) in our country that are linked with each other and with the central information network having access to international database.

World Wide Web: A vast quantum of current data is available on World Wide Web. One of the most important on-line learning centre with power web is www. mhhe. com/ environmental science and multimedia Digital Content Manager (DCM) in the form of CD-ROM provides the most current and relevant information on principles of environmental science, various problems, queries, applications and solutions.

The World Wide Web with resource material on every aspect, class-room activities, digital files of photos, power-point lecture presentations, animations, web-exercises and quiz has proved to be extremely useful both for the students and the teachers of environmental studies.

The role of online learning centre website has the following features:

(a) *Student Friendly Features:* These include practice quiz, how to study tips, hyperlinks on every chapter topics with detailed information, web exercises, case studies, environment maps, key-terms, career information, current articles, interactive encyclopaedia and how to contact your elected officials.

(b) Teacher-friendly features include in addition to above supplement resource charts, additional case studies, answers to web exercises, solutions to critical thinking questions, editing facility to add or delete questions and create multiple versions of same test, etc.

Information technology is expanding rapidly with increasing applications and new avenues are being opened with effective role in education, management and planning in the field of environment and health.

Rehabilitation Programme

Problems and Concerns: Economic development raises the quality and standard of living of the people of a country. Developmental projects are planned to bring benefits to the society. However, in the process of development, very often there is over-exploitation of natural resources and degradation of the environment. Besides this, quite often, the native people of the project site are directly affected. These native people are generally the poorest of the poor, underprivileged tribal people. Various types of projects result in the displacement of the native people who undergo tremendous economic and psychological distress, as the socio-economic and ecological base of the local community is disturbed.

(a) *Displacement Problems due to Dams:* The big river valley projects have one of the most serious socio-economic impacts due to large-scale displacement of local people from their ancestral home and loss of their traditional profession or occupation. India is one of countries in the world leading in big dam construction and in the last 50 years more than 20 million people are estimated to have been directly or indirectly affected by these dams.

The Hirakund Dam has displaced more than 20,000 people residing in about 250 villages. The Bhakra Nangal Dam was constructed during 1950's and till now it has not been possible to rehabilitate even half of the displaced persons.

Same is the case with Tehri Dam on the river Bhagirathi, construction of which was green-signalled after three decades of long campaign against the project by the noted activist Sunder lal Bahuguna, the propagator of Chipko Movement. The immediate impact of the Tehri Dam would be on the 10,000 residents of the Tehri town. While displacement is looming large over the people, rehabilitation has become a more burning issue.

(b) *Displacement due to Mining:* Mining is another developmental activity, which causes displacement of the native people. Several thousands of hectares of land area is covered in mining operation and the native people are displaced. Sometimes displacement of local people is due to accidents occurring in mined areas like subsidence of land that often leads to shifting of people.

(c) *Displacement due to Creation of National Parks:* When some forest area is covered under a National Park, it is a welcome step for conservation of the natural resources. However, it also has a social aspect associated with it which is often neglected. A major portion of the forest is declared as core area, where the entry of local dwellers or tribals is prohibited. When these villagers are deprived of their ancestral right or access to the forests, they usually retaliate by starting destructive activities. There is a need to look into their problems and provide them some employment.

The United Nations Universal Declaration on Human Rights [Article 25(1)] has declared that right to housing is a basic human right.

In India, most of the displacements have resulted due to land acquisition by the government for various reasons. For this purpose, the government has the Land Acquisition Act, 1894 which empowers it to serve notice to the people to vacate their lands if there is a need as per government planning. Provision of cash compensation in lieu of the land vacated exists in section 16 of the Act. The major issues related to displacement and rehabilitation are as follows:

(i) Tribals are usually the most affected amongst the displaced who are already poor. Displacement further increases their poverty due to loss of land, home, jobs, food insecurity, loss of access to common property assets, increased morbidity and mortality and social isolation.

(ii) Break up of families is an important social issue arising due to displacement in which the women are the worst affected and they are not even given cash/land compensation.

(iii) The tribals are not familiar with the market policies and trends. Even if they get cash compensation, they get alienated in the modern economic setup.

(iv) The land acquisition laws ignore the communal ownership of property, which is an inbuilt system amongst the tribals. Thus the tribals lose their communitarian basis of economic and cultural existence. They feel like fish out of water.

(v) Kinship systems, marriages, social and cultural functions, their folk-songs, dances and activities vanish with their displacement. Even when they are resettled, it is individual-based resettlement, which totally ignores communal settlement.

(vi) Loss of identity and loss of the intimate link between the people and the environment is one of the biggest loss. The age-long indigenous knowledge, which has been inherited and experienced by them about the flora, fauna, their uses, etc. gets lost.

There is a need for a comprehensive National Rehabilitation Policy. Different states are following different practices in this regard.

There is a need to raise public awareness on these issues to bring the resettlement and rehabilitation plans on a humane footing and to honour the human rights of the oustees.

Ethics of Environment—Issues and Solutions Possible: Environmental ethics refers to the issues, principles and guidelines relating to human interactions with their environment. It is rightly said, "The environmental crisis is an outward manifestation of the crisis of mind and spirit". It all depends on how do we think and act. If we think "Man is all powerful and the supreme creature on this earth and man is the master of nature and can harness it at his will", it reflects our human-centric thinking. On the other hand, if we think "Nature has provided us with all the resources for leading a beautiful life and she nourishes us like a mother, we should respect her and nurture her", this is an earth-centric thinking.

The first view urges us to march ahead gloriously to conquer the nature and establish our supremacy over nature through technological innovations, economic growth and development without much botheration to care for the damage done to the planet earth. The second view urges us to live on this earth as a part of it, like any other creation of Nature and live sustainably. So, we can see that our acts will follow what we think. If we want to check the environmental crisis, we will have to transform our thinking and attitude. That in turn, would transform our deeds, leading to a better environment and better future.

These two world-views are discussed here in relation to environmental protection:

(a) *Anthropocentric World View:* This view is guiding most industrial societies. It puts human beings in the centre giving them the highest status. Man is considered to be most capable for managing the planet earth. The guiding principles of this view are:

(i) Man is the planet's most important species and is the in-charge of the rest of nature.

(ii) Earth has an unlimited supply of resources and it all belongs to us.

(iii) Economic growth is very good and more the growth, the better it is, because it raises our quality of life and the potential for economic growth is unlimited.

(iv) A healthy environment depends upon a healthy economy.

The success of mankind depends upon how good managers we are for deriving benefits for us from nature.

(b) *Eco-centric World View:* This is based on earth-wisdom. The basic beliefs are as follows:

(i) Nature exists not for human beings alone, but for all the species.

(ii) The earth resources are limited and they do not belong only to human beings.

(iii) Economic growth is good till it encourages earth-sustaining development and discourages earth-degrading development.

(iv) A healthy economy depends upon a healthy environment.

(v) The success of mankind depends upon how best we can cooperate with the rest of the nature while trying to use the resources of nature for our benefit.

Environmental ethics can provide us the guidelines for putting our beliefs into action and help us decide what to do when faced with crucial situations. Some important ethical guidelines known as Earth ethics or Environmental Ethics are as follows:

- You should love and honour the earth since it has blessed you with life and governs your survival.
- You should keep each day sacred to earth and celebrate the turning of its seasons.
- You should not hold yourself above other living things and have no right to drive them to extinction.
- You should be grateful to the plants and animals which nourish you by giving you food.
- You should limit your offsprings because too many people will overburden the earth.
- You should not waste your resources on destructive weapons.
- You should not run after gains at the cost of nature, rather should strive to restore its damaged majesty.
- You should not conceal from others the effects you have caused by your actions on earth.
- You should not steal from future generations their right to live in a clean and safe planet by impoverishing or polluting it.

- You should consume the material goods in moderate amounts so that all may share the earth's precious treasure of resources.

If we critically go through the above ten commandments for earth ethics and reflect upon the same, we will find that various religions teach us the same things in one form or the other. Our Vedas have glorified each and every component of nature as gods or goddesses so that people have a feeling of reverence for them. Our religious and cultural rituals make us perform such actions that would help in the conservation of nature and natural resources. The concept of 'ahimsa' (non-violence) in Buddhism and Jainism ensure the protection and conservation of all forms of life, thereby keeping the ecological balance of the earth intact. Our teachings on "having fewer wants" ensures to put "limits to growth" and thus, guide us to have an eco-centric life-style.

Change of Climate

Climate is the average weather of an area. It is the general weather conditions, seasonal variations and extremes of weather in a region. Such conditions which average over a long period-at least 30 years is called climate. The Intergovernmental Panel on Climate Change (IPCC) in 1990 and 1992 published best available evidence about past climate change, the green house effect and recent changes in global temperature. It is observed that earth's temperature has changed considerably during the geological times. It has experienced several glacial and inter-glacial periods. However, during the past 10,000 years of the current interglacial period the mean average temperature has fluctuated by 0.5-1°C over 100 to 200 year period. We have relatively stable climate for thousands of years due to which we have practised agriculture and increased in population. Even small changes in climatic conditions may disturb agriculture that would lead to migration of animals including humans.

Anthropogenic (man-made) activities are upsetting the delicate balance that has established between various components of the environment. Green house gases are increasing in the atmosphere resulting in increase in the average global temperature.

This may upset the hydrological cycle, result in floods and droughts in different regions of the world, cause sea level rise, changes in agriculture productivity, famines and death of humans as well as livestock.

The global change in temperature will not be uniform everywhere and will fluctuate in different regions. The places at higher latitudes will be warmed up more during late autumn and winter than the places in tropics. Poles may experience 2 to 3 times more warming than the global average, while warming in the tropics may be only 50 to 100% on an average. The increased warming at poles will reduce the thermal gradient between the equator and high latitude regions decreasing the energy available to the heat engine that drives the global weather machine.

This will disturb the global pattern of winds and ocean currents as well as the timing and distribution of rainfall. Shifting of ocean currents may change the climate of Iceland and Britain and may result in cooling at a time when rest of the world warms. By a temperature increase of 1.5 to 4.5°C the global hydrological cycle is expected to intensify by 5 to 10%. Disturbed rainfall will result in some areas becoming wetter and the others drier. Although rainfall may increase, higher temperatures will result in more evapo-transpiration leading to annual water deficit in crop fields.

Global Warming: Troposphere, the lowermost layer of the atmosphere, traps heat by a natural process due to the presence of certain gases. This effect is called Green House Effect as it is similar to the warming effect observed in the horticultural green house made of glass. The amount of heat trapped in the atmosphere depends mostly on the concentrations of "heat trapping" or "green house" gases and the length of time they stay in the atmosphere.

The major green house gases are carbon dioxide, ozone, methane, nitrous oxide, chlorofluorocarbons (CFCs) and water vapours. The average global temperature is 15°C. In the absence of green house gases this temperature would have been 18°C. Therefore, Green House Effect contributes a temperature rise to the tune of 33°C. Heat trapped by green house gases in the atmosphere keeps the planet warm enough to allow us and other species to exist. The two predominant green house gases are water vapours, which are controlled by hydrological cycle, and carbon dioxide, which is controlled mostly by the global carbon cycle. While the levels of water vapour in the troposphere have relatively remained constant, the levels of carbon dioxide have increased. Other gases whose levels have increased due to human activities are methane, nitrous oxide and chlorofluorocarbons. Deforestation has further resulted in elevated levels of carbon dioxide due to non-removal of carbon dioxide by plants through photosynthesis. Warming or cooling by more than 2°C over the past few decades may prove to be disastrous for various ecosystems on the earth including humans, as it would alter the conditions faster than some species could adapt or migrate. Some areas will become inhabitable because of drought or floods following a rise in average sea level.

Greenhouse Gases: The phenomenon that worries the environmental scientists is that due to anthropogenic activities there is an increase in the concentration of the greenhouse gases in the air that absorb infra-red light containing heat and results in the re-radiation of even more of the outgoing thermal infra-red energy, thereby increasing the average surface temperature beyond 15°C. The phenomenon is referred to as the enhanced green house effect to distinguish its effect from the one that has been operating naturally for millennia.

The greenhouse gases present in the troposphere and resulting in an increase in the temperature of air and the earth are discussed here:

Carbon Dioxide: It contributes about 55% to global warming from green house gases produced by human activity. Industrial countries account for about 76% of annual emissions. The main sources are fossil fuel burning (67%) and deforestation, other forms of land clearing and burning (33%). CO_2 stays in the atmosphere for about 500 years. CO_2 concentration in the atmosphere was 355 ppm in 1990 that is increasing at a rate of 1.5 ppm every year.

Chlorofluorocarbons (CFCs): These are believed to be responsible for 24% of the human contribution to greenhouse gases. They also deplete ozone in the stratosphere.

The main sources of CFCs include leaking air conditioners and refrigerators, evaporation of industrial solvents, production of plastic foams, aerosols, propellants, etc. CFCs take 10-15 years to reach the stratosphere and generally trap 1500 to 7000 times more heat per molecule than CO_2 while they are in the troposphere.

This heating effect in the troposphere may be partially offset by the cooling caused when CFCs deplete ozone during their 65 to 110 years stay in the stratosphere. Atmospheric concentration of CFC is 0.00225 ppm that is increasing at a rate of 0.5% annually.

Methane (CH_4): It accounts for 18% of the increased greenhouse gases. Methane is produced when bacteria break down dead organic matter in moist places that lack oxygen such as swamps, natural wetlands, paddy fields, landfills and digestive tracts of cattle, sheep and termites. Production and use of oil and natural gas and incomplete burning of organic material are also significant sources of methane. Methane stays in the atmosphere for 7-10 years. Each methane molecule traps about 25 times as much heat as a CO_2 molecule. Atmospheric concentration of methane is 1.675 ppm and it is increasing at a rate of 1% annually.

Nitrous Oxide (N_2O): It is responsible for 6% of the human input of green house gases. Besides trapping heat in the troposphere it also depletes ozone in the stratosphere. It is released from nylon products, from burning of biomass and nitrogen-rich fuels (especially coal) and from the breakdown of nitrogen fertilizers in soil, livestock wastes and nitrate contaminated ground water. Its lifespan in the troposphere is 140-190 years and it traps about 230 times as much heat per molecule as CO_2. The atmospheric concentration of N_2O is 0.3 ppm and is increasing at a rate of 0.2% annually.

Aftermath of Enhanced Greenhouse Effect: The enhanced greenhouse effect will not only cause global warming but will also affect various other climatic and natural processes.

(i) *Global Temperature Increase:* It is estimated that the earth's mean temperature will rise between 1.5 to 5.5°C by 2050 if input of greenhouse gases continues to rise at the present rate. Even at the lower value, earth would be warmer than it has been for 10,000 years.

(ii) *Rise in Sea Level:* With the increase in global temperature sea water will expand. Heating will melt the polar ice sheets and glaciers resulting in further rise in sea level. Current models indicate that an increase in the average atmospheric temperature of 3°C would raise the average global sea level by 0.2-1.5 meters over the next 50-100 years.

One meter rise in sea level will inundate low lying areas of cities like Shanghai, Cairo, Bangkok, Sydney, Hamburg and Venice as well as agricultural lowlands and deltas in Egypt, Bangladesh, India, China and will affect rice productivity. This will also disturb many commercially important spawning grounds, and would probably increase the frequency of storm damage to lagoons, estuaries and coral reefs.

In India, the Lakshadweep Islands with a maximum height of 4 meters above the level may be vulnerable. Some of the most beautiful cities like Mumbai may be saved by heavy investment on embankment to prevent inundation.

Life of millions of people will be affected, by the sea level rise who have built homes in the deltas of the Ganges, the Nile, the Mekong, the Yangtze and the Mississippi rivers.

(iii) *Effects on Human Health:* The global warming will lead to changes in the rainfall pattern in many areas, thereby affecting the distribution of vector-borne diseases like malaria, filariasis, elephantiasis, etc.

Areas which are presently free from diseases like malaria, schistosomiasis, etc. may become the breeding grounds for the vectors of such diseases. The areas likely to be affected in this manner are Ethiopia, Kenya and Indonesia. Warmer temperature and more water stagnation would favour the breeding of mosquitoes, snails and some insects, which are the vectors of such diseases.

Higher temperature and humidity will increase/ aggravate respiratory and skin diseases.

(iv) *Effects on Agriculture:* There are different views regarding the effect of global warming on agriculture. It may show positive or negative effects on various types of crops in different regions of the world. Tropical and subtropical regions will be more affected since the average temperature in these regions is already on the higher side. Even a rise of 2°C may be quite harmful to crops. Soil moisture will decrease and evapotranspiration will increase, which may drastically affect wheat and maize production.

Increase in temperature and humidity will increase pest growth like the growth of vectors for various diseases. Pests will adapt to such changes better than the crops.

To cope up with the changing situation drought resistant, heat resistant and pest resistant varieties of crops have to be developed.

Depletion of Global Warming

To slow down enhanced global warming the following steps will be important:

(i) Cut down the current rate of use of CFCs and fossil fuel.

(ii) Use energy more efficiently.

(iii) Shift to renewable energy resources.

(iv) Increase Nuclear Power Plants for electricity production.

(v) Shift from coal to natural gas.

(vi) Trap and use methane as a fuel.

(vii) Reduce beef production.

(viii) Adopt sustainable agriculture.

(ix) Stabilise population growth.

(x) Efficiently remove CO_2 from smoke stacks.

(xi) Plant more trees.

(xii) Remove atmospheric CO_2 by utilizing photosynthetic algae.

Acid Rain: Oxides of sulfur and nitrogen originating from industrial operations and fossil fuel combustion are the major sources of acid forming gases. Acid forming gases are oxidised over several days by which time they travel several thousand kilometers. In the atmosphere these gases are ultimately converted into sulfuric and nitric acids. Hydrogen chloride emission forms hydrochloric acid. These acids cause acidic rain. Acid rain is only one component of acidic deposition. Acidic deposition is the total of wet acidic deposition (acid rain) and dry deposition.

Rain water is turned acidic when its pH falls below 5.6. In fact clean or natural rain water has a pH of 5.6 at 20°C because of formation of carbonic acid due to dissolution of CO, in water.

The Adirondack Lakes located in the state of New York are known to receive acid rains.

The strong acids like sulphuric acid (H_2SO_4) and nitric acid (HNO_3) dissolved or formed in rainwater dissociate or release hydrogen ions thereby increasing the acidity in rain drops.

Generally sulfuric acid forms a major fraction of acid rain, followed by nitric acid and a very small fraction of other acids. However, in urban areas calcium (Ca^{2+}), Magnesium (Mg^{2+}) and ammonium (NH^{4+}) ions help to neutralise the rain drops shifting the overall H^+ towards basic scale. The overall pH of any raindrop is due to the net effect of carbonic acid, sulfuric acid, nitric acid and other acidic constituents or any neutralisers such as ammonia.

In the absence of rain, dry deposition of acid may occur. Acid-forming gases like oxides of sulphur and nitrogen and acid aerosols get deposited on the surface of water bodies, vegetation, soil and other materials. On moist surfaces or in liquids these acid forming gases can dissolve and form acids similar to that formed in acid rain. If the oxidisers are present on the liquid surfaces then these gases undergo oxidation to form acids. Fine particles or acid droplets can act as nuclei for water to condense to form rain droplets. By such process sulfuric acid is incorporated into the droplets. In the clouds additional SO_2 and NO_2 contact the droplets and get absorbed which can be oxidised by the dissolved hydrogen peroxide (H_2O_2) or other oxidisers.

In the droplets falling from the clouds additional acidic gases and aerosol particles get incorporated, further decreasing their pH. A unit decrease in pH value causes 10 times increase in acidity. Average pH in rainfall over eastern United States from April 1979 to March 1980 was less than 5.0. In India acid rain is recorded from certain places:

Name of place	pH of rainwater
Kodaikanal	5.18
Minicoy	5.52
Mohanbari	5.50

Damaging Impact of Acid Rain: Acid rain causes a number of harmful effects below pH 5.1. The effects are visible in the aquatic system even at pH less than 5.5.

- It causes deterioration of buildings especially made of marble *e.g.* monuments like Taj Mahal. Crystals of calcium and magnesium sulphate are formed as a result of corrosion caused by acid rain.
- It damages stone statues. Priceless stone statues in Greece and Italy have been partially dissolved by acid rain.
- It damages metals and car finishes.
- Aquatic life especially fish are badly affected by lake acidification.
- Aquatic animals suffer from toxicity of metals such as aluminium, mercury, manganese, zinc and lead which leak from the surrounding rocks due to acid rain.
- It results in reproductive failure, and killing of fish.

- Many lakes of Sweden, Norway, Canada have become fishless due to acid rain.
- It damages foliage and weakens trees.
- It makes trees more susceptible to stresses like cold temperature, drought, etc. Many insects and fungi are more tolerant to acidic conditions and hence they can attack the susceptible trees and cause diseases.
- Control of Acid Rain
- Emission of SO_2 and NO_2 from industries and power plants should be reduced by using pollution control equipments.
- Liming of lakes and soils should be done to correct the adverse effects of acid rain.
- A coating of protective layer of inert polymer should be given in the interior of water pipes for drinking water.

9

Social Factors

Human beings live in both natural and social world. Our technological development has strong impacts on the natural as well as the social components. When we talk of development, it cannot be perceived as development only for a privileged few who would have a high standard of living and would derive all the benefits. Development also does not mean an increase in the GNP (Gross National Product) of a few affluent nations. Development has to be visualised in a holistic manner, where it brings benefits to all, not only for the present generation, but also for the future generations. There is an urgent need to inter-link the social aspects with development and environment. In this unit we shall discuss various social issues in relation to environment.

Development from Unsustainable to Sustainable Level: Sustainable development is defined as "meeting the needs of the present without compromising the ability of future generations to meet their own needs." This definition was given by the Norwegian Prime Minister, G.H. Brundtland, who was also the Director of World Health Organisation (WHO). Today sustainable development has become a buzz word and hundreds of programmes have been initiated in the name of sustainable development.

If you want to test whether or not a proposal will achieve the goals of sustainability just try to find out the following. Does it protect our biodiversity? Does it prevent soil erosion? Does it slow down population growth? Does it increase forest cover? Does it cut off the emissions of CFC, SOx, NOx and CO_2? Does it reduce waste generation and does it bring benefits to all? These are only a few parameters for achieving sustainable growth.

Until now development has been human-oriented, that too mainly, for a few rich nations. They have touched the greatest heights of scientific and technological development, but at what cost? The air we breathe, the water we drink and the food we eat have all been badly polluted. Our natural resources are just dwindling due to over exploitation. If growth continues in the same way, very soon we will be facing a "doom's day" as suggested by Meadows *et al* (1972) in their world-famous academic report "The Limits to Growth". This is unsustainable development which will lead to a collapse of the inter-related systems of this earth.

Although the fears about such unsustainable growth and development started in 1970's, yet a clear discussion on sustainable development emerged on an international level in 1992, in the UN Conference on Environment and Development (UNCED), popularly known as The Earth Summit, held at Rio de Janeiro, Brazil. The Rio Declaration aims at "a new and equitable global partnership through the creation of new levels of cooperation among states...." Out of its five significant agreements Agenda-21 proposes a global programme of action on sustainable development in social, economic and political context for the 21st Century.

These are the key aspects for sustainable development:

(a) *Inter-generational Equity:* This emphasises that we should minimise any adverse impacts on resources and environment for future generations, i.e. we should hand over a safe, healthy and resourceful environment to our future generations. This can be possible only if we stop over-exploitation of resources, reduce waste discharge and emissions and maintain ecological balance.

(b) *Intra-generational Equity:* This emphasises that the development processes should seek to minimise the wealth gaps within and between nations. The Human Development Report of United Nations (2001) emphasises that the benefits of technology should seek to achieve the goals of intra-generational equity. The technology should address to the problems of the developing countries, producing drought-tolerant varieties for uncertain climates, vaccines for infectious diseases, clean fuels for domestic and industrial use. This type of technological development will support the economic growth of the poor countries and help in narrowing the wealth gap and lead to sustainability.

Measures for Sustainable Development: Some of the important measures for sustainable development are as follows:

- Using appropriate technology is one which is locally adaptable, eco-friendly, resource-efficient and culturally suitable. It mostly involves local resources and local labour. Indigenous technologies are more useful, cost-effective and sustainable.

  Nature is often taken as a model, using the natural conditions of that region as its components. This concept is known as "design with nature".

  The Technology should use less of resources and should produce minimum waste.

- *Reduce, Reuse, Recycle Approach:* The 3-R approach advocating minimisation of resource use, using them again and again instead of passing it on to the waste stream and recycling the materials goes a long way in achieving the goals of sustainability. It reduces pressure on our resources as well as reduces waste generation and pollution.

- *Prompting Environmental Education and Awareness:* Making environmental education the centre of all learning process will greatly help in changing the thinking and attitude of people towards our earth and the environment. Introducing the subject right from the school stage will inculcate a feeling of belongingness to earth in the small children. 'Earth thinking' will gradually get incorporated in our thinking and action which will greatly help in transforming our life-styles to sustainable ones.
- *Resource Utilisation as per Carrying Capacity:* Any system can sustain a limited number of organisms on a long-term basis which is known as its carrying capacity. In case of human beings, the carrying capacity concept becomes all the more complex. It is because unlike other animals, human beings, not only need food to live, but need so many other things to maintain the quality of life.

Sustainability of a system depends largely upon the carrying capacity of the system. If the carrying capacity of a system is crossed (say, by over exploitation of a resource), environmental degradation starts and continues till it reaches a point of no return.

Carrying capacity has two basic components:

- Supporting capacity, i.e. the capacity to regenerate
- Assimilative capacity, i.e. the capacity to tolerate different stresses.

In order to attain sustainability it is very important to utilise the resources based upon the above two properties of the system. Consumption should not exceed regeneration and changes should not be allowed to occur beyond the tolerance capacity of the system.

Indian Scenario

India has still to go a long way in implementing the concept of sustainable development. We have to lay emphasis on framing a well-planned strategy for our developmental activities while increasing our economic growth. We have tremendous natural diversity as well as a huge population which makes planning for sustainable growth all the more important and complex.

The National Council of Environmental Planning and Coordination (NCPC) set up in 1972 was the focal agency in this regard. The Ministry of Environment & Forests, set up in 1985 has formulated guidelines for various developmental activities keeping in view the sustainability principles.

Urban Problems Connected with Energy: Cities are the main centres of economic growth, trade, education, innovations and employment. Until recently, a big majority of human population lived in rural areas and their economic activities centered around agriculture, cattle rearing, fishing, hunting or some cottage industry.

It was some 200 years ago, with the dawn of Industrial era, the cities showed a rapid development. Now about 50 per cent of the world population lives in urban areas and there is increasing movement of rural folk to cities in search of employment. The urban growth is so fast that it is becoming difficult to accommodate all the industrial, commercial and residential facilities within a limited municipal boundary. As a result, there is spreading of the cities into the sub-urban or rural areas too, a phenomenon known as *urban sprawl.* In developing countries too urban growth is very fast and in most of the cases it is uncontrollable and unplanned growth. In contrast to the rural setup the urban setup is densely populated, consumes a lot of energy and materials and generates a lot of waste.

The energy requirements of urban population are much higher than that of rural ones. This is because urban people have a higher standard of life and their life-style demands more energy inputs in every sphere of life. The energy-demanding activities include:

(i) Residential and commercial lighting.

(ii) Transportation means including automobiles and public transport for moving from residence to workplace.

(iii) Modern life-style using a large number of electrical gadgets in day-to-day life.

(iv) Industrial plants using a big proportion of energy.

(v) A large amount of waste generation which has to be disposed off properly using energy-based techniques.

(vi) Control and prevention of air and water pollution which need energy-dependent technologies.

Due to high population density and high energy demanding activities, the urban problems related to energy are much more magnified as compared to the rural population.

Saving of Water

Water being one of the most precious and indispensable resources needs to be conserved. The following strategies can be adopted for conservation of water.

(i) *Decreasing Runoff Losses:* Huge water-loss occurs due to runoff on most of the soils, which can be reduced by allowing most of the water to infiltrate into the soil. This can be achieved by using contour cultivation, terrace farming, water spreading, chemical treatment or improved water-storage system.

- Contour cultivation on small furrows and ridges across the slopes trap rainwater and allow more time for infiltration. Terracing constructed on deep soils have large water-storage capacity. On gentle slopes trapped run off is spread over a large area for better infiltration.
- Conservation-bench terracing involves construction of a series of benches for catching the run off water.
- Water spreading is done by channeling or lagoon-levelling. In channeling, the water-flow is controlled by a series of diversions with vertical intervals. In lagoon leveling, small depressions are dug in the area so that there is temporary storage of water.
- Chemical-wetting agents (Surfactants) increase the water intake rates when added to normal irrigated soils.
- Surface crop residues, tillage, mulch, animal residues, etc. help in reducing runoff by allowing more time for water to penetrate into the land.
- Chemical conditioners like gypsum ($CaSO_4.2H_2O$) when applied to sodic soils improve soil permeability and reduce run off. Another useful conditioner is HPAN (hydrolysed polyacrylonitrile).
- Water-storage structures like farm ponds, dugouts, etc. built by individual farmers can be useful measures for conserving water through reduction of runoff.

(ii) *Reducing Evaporation Losses:* This is more relevant in humid regions. Horizontal barriers of asphalt placed below the soil surface increase water availability and increase crop yield by 35-40%. This is more effective on sandy soil but less effective on loamy sand soils.

A co-polymer of starch and acrylonitrile called 'super slurper' has been reported to absorb water upto 1400 times its weight. The chemical has been found to be useful for sandy soils.

(iii) *Storing Water in Soil:* Storage of water takes place in the soil root zone in humid regions when the soil is wetted to field capacity. By leaving the soil fallow for one season water can be made available for the crop grown in next season.

(iv) *Reducing Irrigation Losses*

- Use of lined or covered canals to reduce seepage.
- Irrigation in early morning or late evening to reduce evaporation losses.
- Sprinkling irrigation and drip irrigation to conserve water by 30-50%.
- Growing hybrid crop varieties with less water requirements and tolerance to saline water help conserve water.

(v) *Re-use of Water*

- Treated waste water can be used for ferti-irrigation.
- Using grey water from washings, bath-tubs, etc. for watering gardens, washing cars or paths help in saving fresh water.

(vi) *Preventing Wastage of Water:* This can be done in households, commercial buildings and public places.

- Closing taps when not in use
- Repairing any leakage from pipes
- Using small capacity flush in toilets.

(vii) *Increasing Block Pricing:* The consumer has to pay a proportionately higher bill with higher use of water. This helps in economic use of water by the consumers.

Conservation of Rainwater

Rainwater harvesting is a technique of increasing the recharge of groundwater by capturing and storing rainwater. This is done by constructing special water-harvesting structures like dug wells, percolation pits, lagoons, check dams, etc. Rainwater, wherever it falls, is captured and pollution of this water is prevented. Rainwater harvesting is not only proving useful for poor and scanty rainfall regions but also for the rich ones.

The annual average rainfall in India is 1200 mm, However, in most places it is concentrated over the rainy season, from June to September. It is an astonishing fact that Cherapunji, the place receiving the second highest annual rainfall as 11000 mm still suffers from water scarcity. The water flows with run off and there is little vegetation to check the run off and allow infiltration. Till now there is hardly any rain-water harvesting being done in this region, thereby losing all the water that comes through rainfall.

Rainwater harvesting has the following objectives:

- (i) to reduce run off loss
- (ii) to avoid flooding of roads
- (iii) to meet the increasing demands of water
- (iv) to raise the water table by recharging ground water
- (v) to reduce groundwater contamination
- (vi) to supplement groundwater supplies during lean season.

Rainwater can be mainly harvested by any one of the following methods:

- (i) by storing in tanks or reservoirs above or below ground.
- (ii) by constructing pits, dug-wells, lagoons, trench or check-dams on small rivulets
- (iii) by recharging the groundwater.

Before adopting a rain-water harvesting system, the soil characteristics, topography, rainfall pattern and climatic conditions should be understood.

Rain Water Harvesting in Traditional Way: In India, it is an old practice in high rainfall areas to collect rainwater from roof-tops into storage tanks. In foot hills, water flowing from springs are collected by embankment type water storage. In Himalayan foot-hills people use the hollow bamboos as pipelines to transport the water of natural springs. Rajasthan is known for its *'tonkas'* (underground tanks) and *khadins* (embankments) for harvesting rainwater.

In our ancient times we had adequate *Talaabs, Baawaris, Johars, Hauz,* etc. in every city, village and capital cities of our kings and lords, which were used to collect rain-water and ensured adequate water supply in dry periods.

Sophisticated Methods for Rain Water Harvesting: In arid and semi-arid regions artificial ground water recharging is done by constructing shallow percolation tanks. Check-dams made of any suitable native material (brush, poles, rocks, plants, loose rocks, wire-nets, stones, slabs, sacks, etc.) are constructed for harvesting runoff from large catchment areas. Rajendra Singh of Rajasthan popularly known as "water man" has been doing a commendable job for harvesting rainwater by building checkdams in Rajasthan and he was honoured with the prestigious Magsaysay Award for his work.

Groundwater flow can be intercepted by building groundwater dams for storing water underground. As compared to surface dams, groundwater dams have several advantages like minimum evaporation loss, reduced chances of contamination, etc.

In roof top rainwater harvesting, which is a low-cost and effective technique for urban houses and buildings, the rain-water from the top of the roofs is diverted to some surface tank or pit through a delivery system which can be later used for several purposes. Also, it can be used to recharge underground aquifers by diverting the stored water to some abandoned dug-well or by using a hand pump.

All the above techniques of rainwater harvesting are low-cost methods with little maintenance expenses. Rainwater harvesting helps in recharging the aquifers, improves groundwater quality by dilution, improves soil moisture and reduces soil erosion by minimizing runoff water.

Watershed Management: The land area drained by a river is known as the river basin. The watershed is defined as the land area from which water drains under gravity to a common drainage channel. Thus, watershed is a delineated area with a well-defined topographic boundary and one water outlet. The watershed can range from a few square kilometers to few thousand square kilometers in size. In the watershed the hydrological conditions are such that water becomes concentrated within a particular location like a river or a reservoir, by which the watershed is drained. The watershed comprises complex interactions of soil, landform, vegetation, land use activities and water. People and animals are an integral part of a watershed having mutual impacts on each other. We may live anywhere, we would be living in some watershed.

A watershed affects us as it is directly involved in sustained food production, water supply for irrigation, power generation, transportation as well as for influencing sedimentation and erosion, vegetation growth, floods and droughts. Thus, management of watersheds, treating them as a basic functional unit, is extremely important and the first such Integrated Watershed Management was adopted in 1949 by the Damodar Valley Corporation.

Watershed Degradation: The watersheds are very often found to be degraded due to uncontrolled, unplanned and unscientific land use activities. Overgrazing, deforestation, mining, construction activities, industrialisation, shifting cultivation, natural and artificial fires, soil erosion and ignorance of local people have been responsible for degradation of various watersheds.

Objectives of Watershed Management: Rational utilisation of land and water resources for optimum production causing minimum damage to the natural resources is known as watershed management. The objectives of watershed management are as follows:

(i) To rehabilitate the watershed through proper land use adopting conservation strategies for minimizing soil erosion and moisture retention so as to ensure good productivity of the land for the farmers.

(ii) To manage the watershed for beneficial developmental activities like domestic water supply, irrigation, hydropower generation, etc.

(iii) To minimise the risks of floods, droughts and landslides.

(iv) To develop rural areas in the region with clear plans for improving the economy of the region.

Techniques of Watershed Management: In the Fifth Five Year Plan, watershed management approach was included with a number of programmes for it and a national policy was developed. In watershed management, the aspects of development are considered with regard to the availability of resources.

The practices of conservation and development of land and water are taken up with respect to their suitability for peoples' benefit as well as sustainability. Various measures taken up for management include the following:

(i) *Water Harvesting:* Proper storage of water is done with provision for use in dry seasons in low rainfall areas. It also helps in moderation of floods.

(ii) *Afforestation and Agro-forestry:* In watershed development, afforestation and crop plantation play a very important role. They help to prevent soil erosion and retention of moisture. In high rainfall areas woody trees are grown in between crops to substantially reduce the runoff and loss of fertile soil. In Dehradun, trees like Eucalyptus and Leucaena and grasses like Chrysopogon are grown along with maize or wheat to achieve the above objectives. Woody trees grown successfully in such agroforestry programmes include Dalbergia sissoo (Sheesham), Tectona grandis (Teak) and Acacia nilotica (Keekar) which have been used in watershed areas of river Yamuna.

(iii) *Mechanical Measures for Reducing Soil Erosion and Runoff Losses:* Several mechanical measures like terracing, bundling, bench terracing, no-till farming, contour cropping, strip cropping, etc. are used to minimise runoff and soil erosion particularly on the slopes of watersheds. Bundling has proved to be a very useful method in reducing runoff, peak discharge and soil loss in Dehradun and Siwaliks.

(iv) *Scientific Mining and Quarrying:* Due to improper mining, the hills lose stability and get disturbed resulting in landslides, rapid erosion, etc. Contour trenching at an interval of 1 meter on overburden dump, planting some soil binding plants like Ipomoea and Vitex and draining of water courses in the mined area are recommended for minimizing the destructive effects of mining in watershed areas.

(v) *Public Participation:* People's involvement including the farmers and tribals is the key to the success of any watershed management programme, particularly the soil and water conservation. People's cooperation as well as participation has to be ensured for the same.

The communities are to be motivated for protecting a freshly-planted area and maintaining a water harvesting structure implemented by the government or some external agency (NGO) independently or by involving the local people. Properly educating the people about the campaign and its benefits or sometimes paying certain incentives to them can help in effective people's participation.

Successful watershed management has been done at Sukhomajri Panchkula, Haryana through active participation of the local people.

Watershed management in Himalayan region is of vital importance since most of the watersheds of our country lie here. Several anthropogenic activities accelerate its slope instability which need to be prevented and efforts should be made to protect the watershed by preventing overgrazing, terracing and contour farming to check runoff and erosion, etc. On steeper slopes with sliding faces, straw mulching tied with thin wires and ropes helps in establishing the vegetation and stabilizing the slopes.

Environmental Hazards

Pollution of the environment is one of the most horrible ecological crisis to which we are subjected today. We know that three basic amenities for living organisms are air, land or soil and water. Some times in the past, these amenities were pure, virgin, undisturbed, uncontaminated and basically most hospitable for living organisms. But situation is just the reverse today, because progress in science and technology is also leading to pollution of environment and serious ecological imbalance which in the long run, may prove disastrous for mankind. Environment pollution is the result of urban-industrial technological revolution and speedy exploitation of every bit of natural resources.

The mad rat race among nations over the entire globe for development jeopardised the existence of man itself. The craze of progress in agriculture, industry, transportation and technology is taken as the general criterion of development of any nation. This craze resulted into unlimited exploitation of every bit of natural resources.

Such activities of man has created adverse effects on all living organisms in the biosphere. Rapid industrialisation has left with us polluted rivers, contaminated soil, depleted wildlife and exhausted natural resources. Today the environment has become foul, contaminated, undesirable, and therefore, harmful for the health of living organisms, including man. The splendid plentifulness of nature is a heritage that should never be spoiled. But the unlimited rapacious exploitation of nature by man has disturbed the delicate ecological balance existing between living and non-living components on the planet earth. The unfavourable conditions created by man himself threatened the survival not only of man himself but also other living organisms. The number of the species likely to become rare, threatened, endangered or near extinction in the Red Data Book of the IUCN is increasing with time.

Today India, which occupies 7th place among the industrialised developing countries of the world, is provided with a good industrial infrastructure in several industries like chemicals, power, nuclear energy, food, petroleum, pesticides, plastics, etc. A rapid progress in atomic and nuclear energy has added a huge amount of radioactive substances in the atmosphere. A number of industrial effluents and emissions, especially toxic gases are spewed into the air daily. Thus the environment is deteriorated to such an extent that it has crossed the critical limit and has become lethal to all organisms, including man. But what has come out of these developments? We find warnings every where as "Air unfit for breathing", "Water unfit for drinking", "Vegetables unfit for eating" and so on.

The environment pollution is thus necessary evil of all man made progress. It is not only the industrialised countries which are threatened with the environmental pollution, but the menace is growing in India too. The developed countries dump lot of effluents into the environment, polluting the whole globe.

The Theory: Pollution refers to any undesirable change in the physical, chemical or biological characteristics of our environment, i.e. air, water and soil which adversely affect humans or other species of our biosphere directly or indirectly. The word pollution is derived from the Latin word Pollutionem (meaning to defile or make dirty). Pollution can also be defined as an undesirable change in physical, chemical or biological characteristics of water, air and soil that may harmfully affect human, animal and the plant life, industrial progress, living conditions and cultural assets.

Pollution should not be confused with contamination which is the presence of harmful organisms or their products causing disease or discomfort.

Functional Elements: Pollution is brought about by the addition of waste products of human activity to the environment. When the waste products are not efficiently assimilated, decomposed or otherwise removed by the natural, biological and physical processes of the biosphere, adverse effects may result as the pollutants accumulate or get converted into more toxic substances. Thus the materials which cause pollution of the environment are called pollutants.

Pollutant can also be defined as undesirable substances which are present in the wrong place, at the wrong time and in the wrong quantity.

Pollution: Various Kinds

A. Pollutants are divided into two categories on the basis of their forms they exist in the environment after their release:

1. *Primary Pollutants:* Those substances emitted directly from an identifiable source. These are present in the same form in which these are added by man e.g. sulphur dioxide, nitrogen oxide, etc.
2. *Secondary Pollutants:* These are substances derived from primary pollutants by chemical reactions so occur in different forms e.g., Primary pollutants such as hydrocarbons and nitrogen oxides, particularly in the environment react in the presence of sunlight to form a group of nitrous compounds like peroxyacetyl nitrate (PAN) as the secondary pollutant.

   Nitrogen peroxide + Hydrocarbons →
   Ozone + peroxylacetyl nitrate

B. On the basis of their existence in nature, pollutants are divided into two categories:

1. *Quantitative Pollutants:* It is a normal component of the environment but becomes pollutant when its concentration crosses a threshold value e.g., carbon monoxide, carbon dioxide, nitrogen oxide.
2. *Qualitative Pollutants:* A pollutant which is normally not a component of environment but is passed into environment through human action, e.g., pesticide, weedicide, fungicide.

C. From ecosystem point of view, the pollutants may be classified as biodegradable and non-biodegradable pollutants.

1. *Biodegradable Pollutants:* These pollutants are easily degraded completely by micro organisms e.g., domestic sewage, heat, etc. However, if these pollutants enter the environment in such large quantities that complete degradation of all cannot take place, then these become non-biodegradable pollutants and thus pollute the environment.

2. *Non-biodegradable Pollutants:* These pollutants are not normally acted upon by microbes or are acted upon very slowly and thereby pollute the environment. Such pollutants are harmful even in low concentration. These pollutants not only accumulate, but are often biologically magnified as they move in biochemical cycles and along food chains. Non-biodegradable pollutants are further of two sub-types:(i) Wastes e.g., glass, plastics, phenolics, aluminium cans (ii) Poisons e.g., radio active substances, pesticides, smog gases, heavy metals like nickel, chromium, mercury, lead or cadmium.

In general pollutants are residues of substances made by us, used by us and even thrown away by us as waste products which pollute the environment in one way or the other. On the basis of type of environment being polluted, we may recognise air pollution, water pollution, land (soil) pollution.

Air Pollution: Air pollution is defined as the occurrence of foreign particles or gases in the atmosphere which are harmful to man vegetation, animals and buildings. Air pollutant is any solid, liquid or gaseous substance present in the atmosphere in such a concentration that it may or become harmful to human beings, other living creatures, human assets, natural resources and environment. Most of the air pollution is natural due to dust storms, marsh gas, respiration of organisms, decay and decomposition, forest fires, spores. Pollen grains, etc. Only 0.05% of the air pollution is man made but its amount is enormous 5×10^8 tonnes.

Sources: The main sources of air pollution in India are:

Automobiles: The exhaust emitted by automobiles is a major source of air pollution. Automobiles release huge quantities of carbon monoxide (Co), hydrocarbons; and nitrous oxide in the air. Every gallon of petrol consumed by automobiles produces 3 lbs of CO and 15 Pbs nitrogen oxide which is sufficient to pollute 800,000 to 2,00,000 c.c. of air. The combustion of petroleum emits various particulate lead compounds. Mumbai residents have a lead content of 16-18 mg/100ml of blood while lead begins to show toxic effect beyond 20 mg/100ml. About 312 tonnes of lead is let out into Delhi's atmosphere every year. The larger lead particles settle down on the soil and pass into food chain. It is estimated that the highest intensity of CO in the atmosphere during peak traffic hours at some points in Kolkata is about 35 parts per million. In Delhi total calculated vehicular pollution has increased from 424 tonnes/day to 865 tonnes/day during 1980-81 to 1986-87.

There are more petrol driven vehicles in Delhi which collectively add about half a million tonnes of lead in the environment every year. This can endanger the people to dead lead poisoning.

Industrial Chimney Wastes: There are a number of industries which are sources of air pollution. The common air pollutants which are discharged into air from industrial chimneys and power houses are CO, CO_2, SO_2, H_2S and hydrocarbons. These gases are produced due to burning of coal and petroleum and by combustion of lignite at thermal power stations. Many chemical industries release hydrochloric acid, chlorine, nitrogen oxide, zinc, lead, arsenic, oxides of copper in addition to SO_2, CO and H_2S.

Petroleum refineries are the major source of gaseous pollutants. The chief gases are SO_2 and NO_2. Mathura based petroleum refinery is posing threat to Taj Mahal in Agra and other monuments at Fatehpur Sikri. A study was made by central pollution control board which identified five major and 22,000 small industries polluting capital's atmosphere. One can imagine the kind of air inhaled by the residents of the localities surrounding these industrial units.

Thermal Power Stations: There are a number of thermal power stations and super thermal power stations in the country. The National Thermal Power Corporation (NTPC)is setting up four mammoth coal powered power stations to augment the energy generation. These plants generate electricity by burning fossil fuels like coal and in some cases oils, release many pollutants such as flyash, SO_2 and other gases and hydrocarbons Table shows various gaseous pollutants from a 200 MW thermal power plant.

Gaseous pollutants from a 200MW thermal power plant (Coal consumed 6.67 x 200 = 1334, i.e. 1400 tonnes a day)

Components	*Emission factor kg/tonne of coal*	*Emitted quantity (tonnes a day)*
Aldehydes	0.0025	0.0035
Carbon monoxide	0.25	0.35
Hydrocarbons	0.10	0.14
NO_2	10.00	14.00
Oxides of Sulphur (0.5% S)	19(s)	13.30
Particulate matter (33% ash)	8(A)	369.60
Ash	2(A)	92.40

(A) - Ash content in coal in percent (S) - Sulphur content in coal in percent.

Burning of Fuels: Fossil fuels provide us energy for many domestic activities. Among fossil fuels coal and oils are major source of energy. Coal on combustion produces CO_2. Incomplete combustion yields CO and a variety of hydrocarbons including methane. Burning of coal produces SO_2 and ash also.

Agriculture: About 60-65% of CO, is produced from burning of forests. About 40% of methane is produced from paddy fields, guts of livestocks and burning of biomass. Crop spraying and dusting for pest and weed control are responsible for emitting organic phosphates, arsenic and lead into air.

Suspended Participate Matter (SPM): It is a major pollutant of air. Dust is generated from sources such as coal dust, cement dust, silica dust (Stone crushing). Cotton dust is main pollutant in Ahmedabad, Surat, etc. In Delhi's atmosphere about 53,000 tonnes of SPM are added annually.

The Solutions

Carbon Dioxide: It is released in the air through combustion of fuel in homes, factories, etc. It is also produced during the process of respiration and volcanic eruptions. The amount of atmospheric carbon dioxide has been continuously rising over the years due to deforestation, burning of fuels without a corresponding increase in plant population. In 1987, alone earth produced 14,600 million metric tonnes of carbon dioxide. Out of this only 63% was absorbed by plants and oceans. The year added 5900 million tonnes of gas into the atmosphere. Consequently CO_2 concentration of atmosphere has been steadily rising. The global temperature is believed to have risen to 1°c during the last fifty years therefore, causing a warming effect (green house effect). Excess of carbon dioxide produces headache and nausea in human beings.

Carbon Monoxide: CO is a colourless, odourless, non-irritating but very poisonous gas. It is produced by incomplete combustion of fuels. It accounts for over 50% of the total air pollutants. Automobiles produce 74% of all man made carbon monoxide. A good amount of CO is also produced naturally by plants and animals. It is estimated that plants alone give out 10^8 tonnes of carbon monoxide out of a total production of 2.54×10^9 tons from various sources. Carbon monoxide has 200 times stronger affinity for haemoglobin as compared to oxygen. It forms Carboxyhaemoglobin. Oxygen carrying capacity of blood is reduced. When in high concentration, causes headache, giddiness, laziness, reduced vision, nervous and cardiovascular disorders and even death. Carbon monoxide is the largest pollutant in Delhi's air and was measured at about 120,349 tonnes in 1991. In 1997, co-emission was about 132 times more than in 1987.

Sulphuric Dioxide: It is produced in large quantity by burning of fossil fuels (coal and oil) in industries, thermal plants, homes, fertilizer industries, etc. and during smelting of metallic ores (e.g., iron, copper, lead, zinc, nickel, etc.). It is also formed from hydrogen sulphide (H_2S) gas formed from natural sources. The gaseous SO, oxidises to SO_3, which on combination with water forms sulphuric acid. Sulphur dioxide accounts about 18% of total air pollutants. Higher concentration of SO_2 causes chlorosis and necrosis of vegetation. Lichens are very sensitive to SO_2 and are completely destroyed. SO_2 also causes respiratory diseases in man (bronchytis, asthma, emphysema and lung fibrosis). It results in discoloration and deterioration of buildings, sculptures, painted surfaces, fabrics, paper, leather, etc. The exhausts from Mathura refinery are a threat to the Taj at Agra. Sulphur dioxide and sulphur trioxide react with water to form sulphurous acid (H_2SO_3) and sulphuric acid (H_2SO_4) respectively. These acids cause irritation of eyes, nose and throat. The highly corrosive H_2SO_4 attacks exposed metal surfaces such as steel rail tracks. The sulphurous and sulphuric acids may precipitate as rain or snow, producing acid rains or acid precipitation. The acid rains destroy crops and wild plants.

Aerosols: These are chemicals which are present in air in the form of vapours or fine mist. Aerosols are used as disinfectants. They are emitted at high altitudes by jet and contain chlorofluorocarbons (CFCs). Chlorofluorocarbons are used as coolants (in refrigerators and air conditioners). CFCs have a long life. They cause green house effect and deplete ozone layer. Their use is being banned by all countries. CFCs are being replaced by hydro-fluorocarbons (HFC) and hydrochlorofluorocarbons (HCFC). Related compounds are Polychlorinated biphenyls (PCB). They pass into food chain, act as poison and impair reproductive ability of the birds. In human beings PCM causes damage to liver, vision, central nervous system and change in pigmentation.

Fluorides: These are released during refinement of aluminium, rock phosphates, etc. Gaseous fluorides cause necrosis, chlorosis and abscission of leaves, etc.

Nitrogen Oxides: They are produced in the atmosphere electro-photo-chemically from nitrogen and oxygen. Biologically they are formed from nitrates and nitrites by denitrifying baçteria. These are also produced due to combustion process of fuel in industries, automobiles, nitrogen fertilizer plants, etc. These include N_2O, NO, NO_2, N_2O_4 and N_2O_5. The nitrogen oxides form about 10% of air pollutants. Nitrogen oxide act on unsaturated hydrocarbon to form peroxyacyl nitrate (PAN). They are also responsible for forming photochemical smog. Alongwith SO_2, nitrogen oxides cause acid rain. In the presence of moisture, nitrogen oxides have a corrosive effect on metals, marble and lime stone. The oxides possess mutagenic properties. They also cause eye irritation, respiratory troubles, blood congestion and dilation of arteries.

Hydrocarbons: These are produced naturally (e.g., marsh gas), evaporation from fuel dumps, incomplete combustion of fossil fuels, automobile exhausts, refineries, agricultural burning, etc. Methane and ethylene are two common gaseous hydrocarbons. Methane is added by anaerobic decay of organic matter.

It is the principle component of gobar gas. Approximately 57,232 tonnes of hydrocarbons are added in Delhi's air annually and from the survey it was found that motor cycles and scooters contribute the maximum amount of hydrocarbons. Benzene is a major constituent of petrol and automobile exhaust. Ethylene aids in plant seen-science, abscission and fruit ripening. Benzene and Benzpyrene are carcinogenic. Hydrocarbons generally cause bronchial constriction, asthma, irritation of eyes, etc. At high levels in absence of oxygen methane may be narcotic on man.

Photo Chemical Oxidants: These are produced photochemically by reaction between nitrogen oxides and hydrocarbons producing secondary pollutants like ozone, peroxyacyl nitrates (PAN), aldehydes and phenols. The secondary pollutants formed due to the photochemical reactions are called photochemical oxidants.

NO_2	hv (light)	No + O
nitrogen	→	nascent
dioxide	→	oxygen
O_2 + O oxygen	→	O_3 ozone
H.C. + O_3 hydrocarbon	→	R. Co_2 + R. CHO Aldehyde
RCO_2 + NO	→	RCO + NO_2
RCO + NO_2 + O_2	→	R. Co: O NO_2 peroxyacyl nitrate

The photochemical oxides badly affect the performance of human. They adversely affect respiratory system.

Bibliography

Abbasi, S.A. : *Wetlands of India : Ecology and Threats,* Discovery Publishing House, New Delhi, 1997.

Adkins, R.M. : *Annual Review of Ecology and Systematics,* A. R., London, 1993.

Akhtar, R. : *Disease Ecology and Health,* Rawat Publications, Jaipur, 2001.

Alison, J. : *Spatial Ecological : Economic Analysis for Wetland Management,* Cambridge, London, 2004.

Anthony, J. : *Handbook of Ecological Restoration,* Cambridge, London, 2002.

Antonovics, Janis : *Integrating Ecology and Evolution in a Spatial Context,* Cambridge, London, 2001.

Banerjee, B.N. : *Environmental Pollution and Bhopal Killings,* Gyan Books, New Delhi, 1987.

Barker, G. : *Physiological Plant Ecology,* Cambridge, London, 2002.

Berger, W.H. : *Abrupt Climatic Change,* D. Radiel Dordrecht, Boston, 1990.

Bhargava, Gopal : *Ecological Politics : Different Dimensions,* Gyan Books, New Delhi, 2002.

——— : *Marine Ecosystems,* Gyan Books, New Delhi, 2001.

Bhusan, Bibhuti : *Social Ecology of Forest Resources,* Gyan Books, New Delhi, 2004.

Billihgs, W.D. : *Plant, Man and the Ecosystem*, MacMillan, London, 1971.

Bookchin, M. : *The Philosophy of Social Ecology*, Rawat Publications, Jaipur, 1996.

Chandra, Ramesh and Aneja, Ritu : *Corporate Governance for Sustainable Environment*, Gyan Books, New Delhi, 2004.

Colin, J. : *Issues in Environmental Economics*, Blackwell, London, 2002.

Crawford, T.J. and Hewitt, G.M. : *Genes in Ecology*, Cambridge, London, 1992.

Crosby, A.W. : *Ecological Imperialism : The Biological Expansion of Europe*, Cambridge, London, 2000.

Dakshinamurthi, C. : *Water Resources of India and their Utilization in Agriculture*, Oxford University Press, New Delhi, 1973.

Dass, Sujata K. : *Environment Challenges and Sustainable Future*, Gyan Books, New Delhi, 2003.

Dasseman, R.F. : *Environmental Conservation*, John Wiley, New York, 1976.

Dave, M. : *Urban Ecology and Levels of Development*, Rawat Publications, Jaipur,

David, Khemchand : *Advanced Dictionary of Ecology and Environment*, Dominant Publishers, New Delhi, 2000.

DeVere, Burton : *Ecology of Fish and Wildlife*, Delmar, London, 1996.

Dickinson, L. : *Ecology and Evolution of Cooperative Breeding in Birds*, Cambridge, London, 2004.

Dodson, S. I. : *Ecology*, Oxford, New York, 1998.

Dutta, Ashish : *Biodiversity and Ecosystem Conservation*, Gyan Books, New Delhi, 2001.

Dwivedi, S.D. : *Naturopathy : For Perfect Health*, Gyan Books, New Delhi, 2002.

Eric, R. Pianka : *Evolutionary Ecology*, Harper Collins, London, 1994.

Fritz, L. Knopf : *Ecology and Conservation of Great Plains Vertebrates,* Springer, London, 1997.

Gaan, Narottam : *Environmental Security Concept and Dimensons,* Gyan Books, New Delhi, 2004.

Gary, A. Klee : *The Coastal Environment : Toward Integrated Coastal and Marine Sanctuary Management,* Prentice Hall, London, 1999.

George, A. Maul : *Climatic Change in the Intra : Americas Sea,* Arnold, New Jersey, 1993.

Gerry Closs, Barbare Downes and Andrew, Boulton : *Freshwater Ecology : A Scientific Introduction,* Blackwell, London, 2004.

Gordon, D. : *World Population Problems,* Park Press, Campaign, 1984.

Grafton, Q. : *Economics of the Environment,* Blackwell, London, 2004.

Gupta, K. : *Energy and Environment in India : A Study of Energy Management,* Gyan Books, New Delhi, 2002.

Hourani, G.F. : *Arab Seafaring in the Indian Ocean and Early Medieval Times,* Princeton University Press, Princeton, 1951.

Husain, M. : *Evolution of Geographical Thought,* Rawat, Jaipur, 1995.

Hussen, A. : *Principles of Environmental Economics,* Routledge, New York, 2004.

Hutchings, J. : *The Ecological Consequences of Environmental Heterogeneity,* Cambridge, London, 2000.

Ian, D. Whyte : *Climating Change and Human Society,* Arnold, New Jersey, 1995.

Ian, F. Spellerberg : *Monitoring Ecological Change,* Cambridge, London, 2005.

James, K. Lein : *Integrated Environmental Planning,* Blackwell, London, 2003.

James, M. Bullock : *Dispersal Ecology,* Cambridge, London, 2002.

John, E.FA : *Evolution and Ecology of Macque Societies*, Cambridge, London, 1996.

John, N. Rayner : *Dynamic Climatology : Basis in Mathematics and Physics*, Blackwell, London, 2001.

Joshi, M.V. : *Theories and Approaches of Environmental Economics*, Atlantic Publishers, New Delhi, 2004.

Kaur, Karanjot : *Green Revolution : Ecological Implications*, Dominant Publishers, New Delhi, 1999.

Khanna, D.R. : *Microbial Ecology : A Study of River Ganga*, Discovery Publishing House, New Delhi, 2004.

Kumar, Chattopadhyay : *The Desperate Delta : Social Ecology of Sunderbans*, Gyan Books, New Delhi, 1995.

Kumar, Shashi : *Human Ecology for Globalization : Human Ecology in Action*, Atlantic Publishers, New Delhi, 2003.

——— : *Human Ecology for Globalization : New Dimensions for Economic Growth*, Atlantic Publishers, New Delhi, 2004.

Madan, Mohan : *Ecology and Development*, Rawat Publications, Jaipur, 2000.

Martin, T.E. : *Ecology and Management of Neotropical Migratory Birds*, Oxford, New York, 1995.

Mash, G. P. : *Man and Nature Physical Geography as Modified by Human Action*, Charles Scriber, New York, 1964.

Max, N. : *The Environmental Revolution*, McGraw-Hill, New York, 1970.

Michael, Moss : *Issues and Perspectives in Landscape Ecology*, Cambridge, London, 2005.

Moore, Francis : *Environment and Society*, Dominant Publishers, New Delhi, 2003.

Mukherjee, Roma : *Environmental Management and Awareness Issues*, Sterling Publishers, New Delhi, 2002.

Muthiah, S. : *An Atlas of India*, Oxford University Press, New York, 1992.

Myers, H. : *Ecology and Control of Introduced Plants,* Cambridge, London, 2003.

Nadarajah, M. : *Culture, Gender and Ecology : Beyond Workerism,* Rawat Publications, Jaipur, 1999.

Nigel, J.R. : *Mountains at Risk : Current Issues in Environmental Studies,* Manohar Publishers, New Delhi, 1995.

Nigel, Webb : *Ecological Reviews,* Cambridge, London, 2005.

Overbeck, J. : *Microbial Ecology of Lake Plubsee: Ecological Studies,* Springer, London, 1994.

Pandey, Mahendra : *Environment Pollutants & Women's Health,* Dominant Publishers, New Delhi, 2003.

Paul, A. : *Introduction to Ecology,* John Wiley and Sons, New York, 1973.

Peter, J. Grubb and John, B. Whittaker : *Toward a More Exact Ecology,* Cambridge, London, 1989.

Puri, G. : *Essential Soil Science : A Clear and Concise Introduction to Soil Science,* Blackwell, London, 2002.

Rampinot, M.R. : *Climate: History, Periodicity and Predictability,* Von Nostrand Reinholt, New York, 1987.

Ravi, V.S. : *The Living Adventure : Nature and Biology,* Manohar Publishers, New Delhi, 2001.

Raza, M. : *Development and Ecology,* Rawat Publications, Jaipur, 2000.

Ring, T. Carde and Jocelyn, G. Millar : *Advances in Insect Chemical Ecology,* Cambridge, London, 2004.

Rob, H.G. Jongman : *Ecological Networks and Greenways,* Cambridge, London, 2004.

Sarah, J. Woodin and Mick, Marquiss : *Ecology of Arctic Environments,* Cambridge, London, 1997.

Sheth, P. : *Environmentalism : Politics, Ecology and Development,* Rawat Publications, Jaipur, 1997.

Shukla, S.K. and Srivastava, P.R. : *Ecology and The Environment,* Commonwealth Publishers, New Delhi, 1992.

Sinha, R.K. : *Eco-Politics and Development in India*, Commonwealth Publishers, New Delhi, 1996.

Stamou, G.P. : *Arthropods of Mediaterranean : Type Ecosystems*, Springer, London, 1998.

Thomas, R. : *Man's Impact on the Environment*, McGraw-Hill, New York, 1988.

Thomson, T. : *Introductory Essay to Flora Indica*, MacMillan, London, 1855.

Trivedi, P.R. and Dudarshan, K.N. : *India Ecology and Environment*, Commonwealth Publishers, New Delhi, 1994.

Usher, Michael : *Ecology, Biodiversity, and Conservation*, Cambridge, London, 2005.

Vayda, A.P. G. : *Man in the Pacific Islands: Essays on Geographical Change in the Pacific*, Oxford University Press, New York, 1972.

Venus, W. : *Kanpur City: A Study in Environmental Pollution*, Tara Book Agency, Varanasi, 1982.

Vogt, K.A. : *Ecosystems : Balancing Science with Management*, Springer, London, 1997.

Walter, G.H. : *Insect Pest Management and Ecological Research*, Cambridge, London, 2003.

Wilfried, B. : *Atmospheric Pollution*, McGraw Hill, New York, 1972.

William, W. : *Integrating Landscape Ecology into Natural Resource Management*, Cambridge, London, 2002.

Yadav, P.R. : *Environmental Ecology*, Discovery Publishing House, New Delhi, 2003.

Yeates, M. : *Plant Ecology*, Harper & Row, New York, 1980.

Zelinsky, W. : *A Study of Ecology*, Oxford University Press, New York, 1970.

Zuckerman, Ben : *Human Population and the Environmental Crisis*, Jones & Bart., New Jersey, 1996.

Index

A

Acid, 13, 14, 15, 17, 18, 19, 26, 27, 39, 42, 43, 47, 68, 175, 177, 178, 179, 180, 181, 182, 216, 219, 271, 272, 273, 274, 293, 296, 297.

Agriculture, 102, 122, 128, 143, 144, 148, 150, 151, 153, 157, 197, 199, 205, 210, 214, 216, 225, 226, 228, 229, 232, 265, 266, 270, 271, 279, 289, 294.

AIDS, 234, 250, 251, 252, 256.

Air Pollution, 24, 36, 40, 42, 43, 161, 164, 219, 242, 292, 293.

Atmosphere, 23, 24, 25, 26, 28, 29, 30, 36, 42, 124, 125, 127, 128, 129, 131, 134, 136, 140, 152, 204, 207, 219, 221, 222, 224, 266, 267, 268, 272, 289, 292, 293, 295, 297.

B

Bacteria, 7, 8, 25, 37, 48, 52, 61, 103, 132, 133, 134, 136, 137, 150, 165, 220, 223, 240, 268, 297.

Bacterial Diseases, 54.

Biodiversity, 200, 202, 217, 246, 249, 257, 276.

Biological Resources, 257.

Biosphere, 30, 124, 128, 139, 140, 289, 290.

C

Cancer, 20, 21, 22, 241, 242, 251, 256.

Carbon Dioxide, 13, 23, 124, 129, 130, 154, 165, 168, 198, 224, 267, 268, 291, 295.

Carbon Monoxide, 170, 291, 292, 295.

Climate, 1, 2, 3, 111, 117, 125, 204, 221, 227, 265, 266.

Coal, 20, 25, 26, 41, 131, 141, 142, 143, 144, 145, 146, 148, 157, 159, 167, 168, 169, 170, 196, 201, 214, 216, 217, 218, 269, 271, 293, 294, 295, 296.

Compressed Natural Gas, 146, 170.

Copper, 13, 42, 99, 128, 137, 160, 174, 191, 201, 215, 216, 217, 218, 293, 296.

Cropping, 46, 114, 115, 121, 150, 151, 152, 208, 209, 211, 287.

Crops, 48, 55, 84, 85, 86, 88, 91, 96, 102, 103, 109, 114, 115, 122, 135, 141, 150, 151, 152, 164, 196, 205, 211, 229, 252, 270, 271, 287, 296.

Cultivation, 81, 82, 86, 88, 89, 122, 135, 199, 226, 280, 281, 286.

D

Deforestation, 112, 113, 117, 145, 150, 199, 200, 201, 202, 207, 208, 218, 225, 226, 227, 229, 230, 257, 267, 268, 285, 295.

Degradation, 27, 111, 117, 165, 198, 200, 230, 236, 237, 246, 257, 259, 278, 285, 286, 291.

Dusters, 99, 100, 102.

E

Earth Summit, 245, 276.

Economic Development, 235, 259.

Education, 238, 243, 244, 246, 247, 248, 249, 250, 253, 255, 256, 259, 278, 279.

Elements, 15, 23, 30, 35, 98, 124, 128, 129, 135, 148, 214, 215, 250, 290.

Energy Plantations, 144, 164.

Energy Sources, 143, 144, 157, 158, 166, 167.

Environment, 3, 20, 21, 32, 34, 45, 51, 64, 95, 96, 103, 119, 124, 125, 128, 129, 131, 139, 140, 142, 147, 156, 203, 217, 224, 230, 240, 241, 242, 245, 246, 248, 249, 252, 253, 255, 256, 257, 259, 262, 263, 264, 266, 275, 276, 277, 278, 279, 288, 289, 290, 291, 292, 293.

F

Fibres, 23, 131.

Fish, 6, 98, 153, 155, 174, 176, 261, 273.

Food Industry, 37.

Forests, 3, 4, 118, 141, 151, 196, 197, 198, 199, 200, 201, 202, 213, 220, 221, 222, 224, 225, 226, 227, 228, 229, 230, 237, 249, 253, 261, 294.

Fresh Water, 152, 153, 154, 205, 206, 282.

G

Gobar Gas, 150, 165, 297.

Grasslands, 4, 118, 161, 237.

Greenery, 227.

Gross National Product, 275.

Groundwater, 33, 117, 204, 205, 206, 218, 219, 283, 284, 285.

H

Harvesting, 52, 82, 85, 86, 120, 142, 159, 283, 284, 285, 286, 288.

Herbs, 121, 220, 222.

Himalayan Region, 230, 288.

HIV, 234, 250, 251, 252, 256.

Human Beings, 153, 204, 215, 237, 263, 275, 278, 292, 295, 297.

Human Health, 22, 39, 103, 240, 241, 242, 256, 270.

Hydrocarbons, 27, 37, 97, 150, 164, 169, 291, 292, 293, 294, 297, 298.

Hydroelectricity, 144.

Hydrogen, 8, 11, 16, 19, 20, 23, 43, 98, 129, 132, 136, 137, 146, 158, 165, 167, 170, 171, 221, 272, 273, 296.

Hydropower, 148, 157, 158, 162, 286.

I

Indian Scenario, 217, 236, 239, 279.

Insects, 67, 77, 81, 82, 83, 84, 85, 86, 87, 88, 89, 91, 94, 95, 96, 97, 103, 109, 224, 270, 274.

Iron, 19, 26, 128, 137, 142, 173, 177, 187, 214, 215, 216, 217, 218, 220, 296.

L

Lakes, 5, 6, 34, 124, 125, 134, 141, 152, 154, 203, 207, 219, 272, 274.

Land Resources, 111, 150, 196.

Landscape, 201, 218, 224.

Landslides, 116, 117, 151, 200, 201, 202, 286, 287.

Leather, 42, 43, 90, 296.

M

Moisture, 1, 2, 32, 33, 42, 43, 47, 115, 118, 127, 152, 198, 222, 240, 270, 285, 286, 287, 297.

Monsoon, 208, 211, 258.

N

National Institute, 257.

National Parks, 260.

National Thermal Power Corporation, 294.

Natural Gas, 131, 141, 142, 143, 145, 146, 157, 159, 167, 169, 170, 268, 271.

Natural Resources, 118, 119, 122, 140, 141, 142, 143, 196, 245, 246, 250, 252, 254, 259, 260, 265, 276, 286, 288, 289, 292.

Nitrogen, 23, 26, 27, 37, 99, 115, 131, 132, 133, 134, 135, 136, 168, 222, 269, 271, 272, 291, 293, 297, 298.

Nuclear Energy, 37, 148, 170, 171, 289.

Nuclear Power, 143, 144, 148, 171, 271.

Nuclear Reactors, 148, 171.

O

Origin, 98, 143, 183, 184, 185, 186, 190, 193.

Oxides, 26, 30, 37, 43, 168, 271, 272, 291, 293, 294, 297, 298.

Oxygen, 6, 16, 23, 25, 27, 30, 31, 124, 129, 130, 132, 137, 165, 167, 173, 197, 203, 214, 221, 222, 224, 268, 295, 297, 298.

P

Pathology, 45, 52, 94.

Pesticides, 37, 92, 103, 122, 156, 241, 289, 292.

Petro-crops, 164.

Plant Ailments, 46, 48.

Plant Diseases, 45, 47, 52, 63.

Plant Protection, 45, 50, 93.

Plants, 3, 4, 5, 7, 26, 27, 37, 41, 43, 45, 46, 47, 49, 50, 51, 61, 62, 63, 64, 65, 66, 67, 68, 77, 78, 80, 81, 84, 86, 87, 89, 92, 93, 97, 104, 105, 106, 108, 117, 119, 120,

121, 124, 125, 126, 127, 128, 129, 132, 133, 134, 135, 136, 140, 142, 144, 146, 148, 149, 150, 152, 162, 163, 164, 165, 167, 168, 169, 170, 173, 175, 198, 202, 204, 214, 221, 222, 226, 228, 229, 264, 267, 271, 274, 284, 287, 294, 295, 296, 297.

Polluted Air, 36, 245.

Pollution, 6, 20, 23, 24, 33, 36, 40, 42, 43, 103, 122, 123, 141, 144, 146, 153, 161, 162, 164, 170, 171, 198, 205, 217, 220, 221, 225, 237, 242, 246, 255, 257, 274, 278, 280, 283, 288, 290, 292, 293.

Protection, 45, 50, 51, 78, 93, 119, 230, 243, 246, 253, 254, 255, 263, 265.

Q

Quadris, 75, 76.

Quality, 33, 34, 62, 82, 86, 87, 88, 90, 115, 156, 177, 195, 210, 212, 216, 222, 237, 246, 247, 259, 263, 278, 285.

Quantity, 13, 36, 62, 81, 149, 194, 195, 211, 215, 232, 290, 294, 296.

R

Rainfall, 1, 2, 4, 5, 32, 83, 88, 112, 113, 114, 152, 198, 200, 204, 206, 207, 208, 224, 227, 266, 270, 273, 283, 284, 286, 287.

Rainwater, 272, 273, 281, 283, 284, 285.

Redox, 9, 10, 11, 13, 15, 16, 18.

Rivers, 34, 124, 125, 127, 134, 141, 148, 152, 154, 155, 162, 168, 191, 205, 207, 208, 210, 224, 228, 249, 270, 289.

S

Science, 31, 45, 63, 239, 256, 258, 288, 297.

Science and Technology, 256, 288.

Season, 3, 36, 61, 71, 73, 75, 76, 95, 113, 194, 195, 282, 283.

Settlements, 123, 151, 200, 214, 231, 257.

Sewage, 6, 34, 153, 155, 163, 291.

Social Forestry, 209, 229.

Social World, 275.

Soil, 3, 4, 25, 26, 33, 36, 47, 50, 51, 52, 53, 62, 102, 111, 112, 113, 114, 115, 116, 117, 118, 121, 125, 126, 127, 132, 133, 134, 135, 140, 141, 142, 149, 150, 151, 152, 168, 196, 198, 200, 201, 204, 218, 221, 222, 224, 225, 226, 227, 237, 252, 257, 269, 270, 272, 280, 281, 282, 284, 285, 286, 287, 288, 289, 290, 292, 293.

Soil Erosion, 112, 113, 114, 115, 117, 118, 150, 198, 200, 218, 224, 225, 227, 276, 285, 286, 287.

Soil Fertility, 115, 116, 151, 152, 200, 227.

Soil Resources, 151.

Solar Cars, 146, 147.

Solar Energy, 124, 141, 143, 144, 146, 147, 158, 159, 160, 161, 164, 196, 221.

Sulphur, 23, 26, 37, 39, 98, 99, 135, 136, 137, 168, 177, 216, 218, 219, 220, 272, 291, 234, 296.

Syndrome, 68, 174, 175, 178, 250.

T

Technology, 31, 77, 93, 142, 144, 157, 220, 247, 256, 259, 277, 288, 289.

Temperature, 1, 2, 6, 7, 8, 24, 25, 33, 42, 88, 121, 126, 149, 161, 163, 164, 166, 190, 203, 204, 221, 222, 240, 265, 266, 267, 268, 269, 270, 271, 274, 295.

Tidal Energy, 141, 144, 149, 158, 162, 196.

Tools, 103, 139, 247.

Toxicity, 41, 68, 273.

Trees, 4, 67, 84, 86, 89, 115, 120, 121, 135, 146, 150, 159, 164, 197, 200, 201, 203, 220, 222, 224, 226, 227, 228, 229, 230, 271, 274, 287.

U

Urban Areas, 34, 229, 272, 279.

V

Vegetables, 48, 55, 122, 179, 180, 181, 182, 289.

Villagers, 260.

Viral Diseases, 54.

Vitex, 287.

Volume, 14, 15, 23, 38, 100, 101.

W

Wastes, 111, 123, 144, 145, 149, 150, 151, 152, 155, 163, 164, 165, 204, 205, 219, 245, 269, 292, 293.

Water Energy, 143.

Water Resources, 152, 153, 196, 204, 206, 207, 208, 209, 257, 286.

Water Ways, 153.

Watershed, 33, 285, 286, 287, 288.

Wealth, 123, 150, 156, 220, 221, 223, 225, 227, 240, 256, 277.

Weapons, 170, 214, 264.

Welfare, 95, 139, 143, 230, 238, 249, 252, 253, 254, 255.

Wildlife, 196, 198, 200, 256, 257, 289.

Wind Energy, 143, 144, 147, 148, 158, 161, 196.

Wood Energy, 147, 150.

Wool, 174.

World Health Organisation, 239, 240, 243, 275.

World Wide Web, 258.

Y

Yellow, 44, 53, 55, 85, 87, 91, 195.

Yellow Vein, 53, 55.

Z

Zinc, 42, 98, 137, 174, 214, 273, 293, 296.

Zones, 5, 6, 7, 33, 201, 204, 257.

Zoology, 91, 93, 94.

□□□